Chemistry for Every Kid

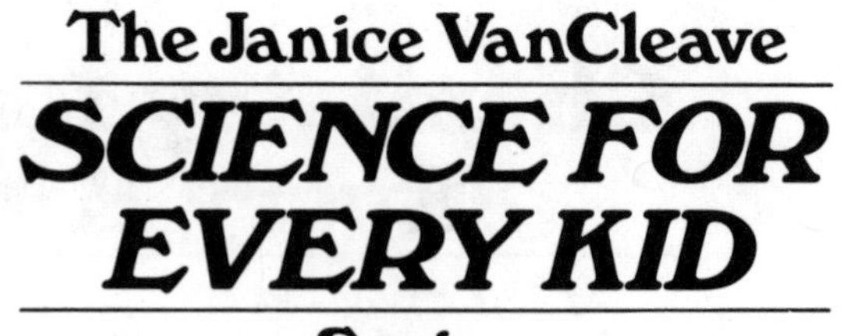
The Janice VanCleave
SCIENCE FOR EVERY KID
Series

Janice VanCleave's
ASTRONOMY
for Every Kid
101 Easy Experiments that Really Work
Exciting ideas, projects, and activities for schools, science fairs, and just plain fun!
Janice VanCleave's
PHYSICS
for Every Kid
101 Easy Experiments in Motion, Heat, Light, Machines, and Sound
Exciting ideas, projects, and activities for schools, science fairs, and just plain fun!

Janice VanCleave's
BIOLOGY
for Every Kid
101 Easy Experiments that Really Work
Exciting ideas, projects, and activities for schools, science fairs, and just plain fun!
Janice VanCleave's
CHEMISTRY
for Every Kid
101 Easy Experiments that Really Work
Exciting ideas, projects, and activities for schools, science fairs, and just plain fun!

Janice VanCleave's
EARTH SCIENCE
for Every Kid
101 Easy Experiments that Really Work
Exciting ideas, projects, and activities for schools, science fairs, and just plain fun!

Chemistry for Every Kid

101 Easy Experiments That Really Work

Janice Pratt VanCleave

John Wiley & Sons, Inc.
New York · Chichester · Brisbane · Toronto · Singapore

Publisher: Stephen Kippur
Editor: David Sobel
Managing Editor: Corinne McCormick
Composition: Vail-Ballou Press, Inc.
Illustrator: April Blair Stewart

Library of Congress Cataloging-in-Publication Data

VanCleave, Janice Pratt.
Chemistry for every kid : 101 easy experiments that really work / by Janice Pratt VanCleave.
p. cm.
Bibliography: p.
Includes index.
Summary: Instructions for experiments, each introducing a different chemistry concept and demonstrating that chemistry is a part of our everyday life.
ISBN 0471-50974-4
ISBN 0-471-62085-8 (pbk.)
1. Chemistry—Experiments. [1. Chemistry—Experiments.
2. Experiments.] I. Title.
QD38.V36 1989
542—dc19 88-27540
CIP
AC

Printed and bound by Courier Companies, Inc.

10 9

Dedicated to my grandchildren
Kimberly, Jennifer, and Davin VanCleave,
and Lauren Russell

Preface

This is a basic science experiment book designed to help children to learn, and adults to teach chemistry concepts, terminology, and laboratory methods. Each experiment was developed to demonstrate that chemistry is a part of our everyday life. All the experiments in this book are unique in their ability to introduce chemistry concepts in a manner that makes learning exciting. These fun-filled experiments will introduce the world of chemistry to young and old alike.

This book contains 101 chemistry experiments. Each experiment has a purpose, list of materials, step-by-step instructions and illustrations, expected results, and a scientific explanation in understandable terms.

The introductory purpose for each experiment gives the reader a clue to the concept that will be introduced. The purpose is complete enough to present the goal, but does not give away the mystery of the results.

Materials are needed, but in all the experiments the necessary items are easily obtained. Most of the materials are readily available around the house. A list of the necessary supplies is given for each experiment.

Detailed step-by-step instructions are given along with illustrations. Pre-testing of all the activities preceded the drafting of the instructions. *The experiments work.*

Expected results are given to direct the experimenter further. They provide immediate positive reinforcement to the student who has performed the experiment properly, and helps correct the student who doesn't achieve the desired results.

Another special feature of the book is the *Why?* section, which gives a scientific explanation for each result in terms that anyone can understand.

This book was written to provide young scientists with safe, workable experiments. The objective of the book is to make the learning of chemistry a rewarding experience and, thus, encourage a student's desire to seek more knowledge about science.

Note:

The experiments and activities in this book should be performed with care and according to the instructions provided. Any person conducting a scientific experiment should read the instructions before beginning the experiment. An adult should supervise young readers who undertake the experiments and activities featured in this book. The publisher accepts no responsibility for any damage caused or sustained while performing the experiments or activities covered by this book.

Acknowledgments

I wish to express my appreciation to my chemistry students at Waco, Texas, High School for testing many of the activities: Selena Baker, Jim Bunting, Yvonne Burton, Diane Craig, Ellen Felder, Kenda Gates, Katie Hensley, Allison High, Curtis Holland, Scharlene Hunt, Daphne Johnson, Dwan Johnson, Steven Johnson, Teresa Jones, Teresa Kirk, Kristy McCollum, Bobby Mentel, Sandra Mental, Meg Miller, Jennifer Ridings, Marcos Rodriquez, Barbara Sanders, Jamey Simmons, Kyle Wagner, Carol Walker, and Tommie Williams.

A special note of gratitude to the members of my family who have volunteered their time to do some of the pre-testing. These special helpers are: Russell, Ginger, Kimberly, Jennifer, David, Tina, and Davin VanCleave, as well as Ginger, Lauren and Calvin Russell.

My deepest gratitude goes to my husband, Wade, who patiently lives in a house filled with jars and bottles in various experimental stages. His encouragement and assistance have been invaluable.

English to SI Substitutions

English	SI (Metric)
LIQUID MEASUREMENTS	
1 gallon	4 liters
1 quart	1 liter
1 pint	500 milliliters
1 cup (8 ounce)	250 milliliters
1 ounce	30 milliliters
1 tablespoon	15 milliliters
1 teaspoon	5 milliliters
LENGTH MEASUREMENTS	
1 yard	1 meter
1 foot (12 inches)	1/3 meter
1 inch	2.54 centimeters
1 mile	1.61 kilometers
PRESSURE	
14.7 pounds per square inch (PSI)	1 atmosphere

Abbreviations

atmosphere = atm
centimeter = cm
cup = c
gallon = gal.
pint = pt.
quart = qt.
ounce = oz.
tablespoon = T.
teaspoon = tsp.
liter = l
milliliter = ml
meter = m
millimeters = mm
kilometers = km
yard = yd.
foot = ft.
inch = in.

Contents

Introduction xvii

1. Matter **1**

1. Kerplunk! 2
2. What's Inside 4
3. Super Chain 6
4. Paper Hop 8
5. Do Not Touch 10
6. Unseen Movement 12
7. An Empty Sack? 14
8. A Rising Ball 16
9. Not at the Same Time 18
10. No Room 20
11. Dry Paper 22
12. How Much? 24
13. Where Did It Go? 26

14. Sinker 28
15. Magic Solution 30

2. Forces 33

16. No Heat 34
17. Rising Water 36
18. Floating Sticks 38
19. Tug of War 40
20. Gravity Won 42
21. Anti-Gravity 44
22. Over the Rim 46
23. Mind of Its Own 48
24. Moving Drop 50
25. Attractive Streams 52
26. Powder Dunk 54
27. Magic Paper 56
28. Spheres of Oil 58
29. Soap Bubbles 60

3. Gases 63

30. Escaping Bubbles 64
31. Foamy Soda 66
32. Pop Cork 68
33. Limewater 70
34. Chemical Breath 72
35. A Hungry Plant 74
36. Erupting Volcano 76
37. How Long? 78
38. Browning Apple 80
39. Disappearing Color 82
40. Fading Color 84
41. Aging Paper 86
42. Rust Prevention 88

4. Changes 91

43. Green Pennies 92
44. Naked Egg 94
45. Breakdown 96
46. Sinking Gel 98
47. Magnesium Milk 100
48. The Green Blob 102
49. Starch I.D. 104
50. Testing for Starch 106
51. Chemical Reactions in Your Mouth 108
52. Magic Writing 110
53. Drinkable Iron 112
54. Curds and Whey 114
55. Limestone Deposits 116
56. A Different Form 118

5. Phase Changes 121

57. Colder Water 122
58. Growing Ice 124
59. Frozen Orange Cubes 126
60. Anti-Freezing 128
61. Chilling Effect 130
62. Crystal Ink 132
63. Fluffy and White 134
64. Frosty Can 136
65. Needles 138
66. Lace 140
67. Cubes 142
68. Plaster Block 144

6. Solutions 147

69. Streamers of Color 148
70. Tasty Solution 150

71. Speedy Soup 152
72. Rainbow Effect 154
73. Falling Snow 156
74. Floating Spheres 158
75. Strengths? 160
76. Spinning 162
77. Layering 164
78. Tyndall Effect 166
79. Immiscible 168
80. Dilution 170
81. Spicy Perfume 172

7. Heat 175

82. Smoke Rings 176
83. Puff Signals 178
84. Clicking Coin 180
85. Chemical Heating 182
86. Heat Changes 184
87. Radiation 186

8. Acid or Base 189

88. Cabbage Indicator 190
89. Cabbage Paper 192
90. Acid–Base Testing 194
91. A or B 196
92. Strong–Stronger 198
93. Drinkable Acid 200
94. Baking with Acid? 202
95. Turmeric Paper 204
96. Now It's Red! 206
97. Wet Only 208
98. Basic Cleaners 210
99. Wood Ash 212

100. Neutral 214
101. Dissolving Fibers 216

Glossary **219**

Resource List **223**

Index **225**

Introduction

Chemistry is the study of the way materials are put together and their behavior under different conditions. This science, more than any other, involves all of one's senses: seeing, hearing, tasting, feeling, and smelling. It is a springboard into other scientific fields. A foundation in basic chemistry facts can assist one in the study of other scientific curricula. One cannot explain the physics concept of magnetism or electricity without understanding the chemistry of atoms. The biological study of photosynthesis has more meaning with the knowledge of the basic chemical reactions involved. Many examples for each scientific field can be given that stress the usefulness of chemistry but, besides this application, chemistry concepts can be used to explain many events we observe in daily life.

The history of chemistry essentially started with the alchemists who tried in vain to transform everything imaginable into gold. All of their efforts to produce gold failed. These so-called "mad scientists" are the forefathers of experimentation. They spent time studying the problem, designed innumerable experiments and, unlike other scientists of their day, they actually experimented. Special equipment was not available for them to order through science catalogs. They designed and made all of the needed flasks and beakers. Some of these designs, in modified form, are still used today in our modern chemistry laboratories.

These chemists of the past were the first to use what is now described as the scientific method, a logical approach to the solution of a problem through experimentation. This process underlies all of the experiments in this book.

The alchemists may have failed to solve their problem of changing a base material into gold, but they set the stage for future scientific discoveries. Their dream has been partially fulfilled after 2000 years. Kenneth T. Bainbridge in 1941 changed mercury into gold. The gold produced did not fill a pot, but rather was microscopic in size, and was more expensive to produce than the gold was worth.

Even today, just the mention of the word chemistry too often conjures up the image of a "mad scientist" hovering over strange, bubbling flasks. Unfortunately, explosions, danger, and flesh-eating acids are associated with chemistry. "Are there any chemicals in there?" is a frequently asked question. This book will relieve the fears of performing chemistry experiments by acquainting the reader with "chemicals." As knowledge about chemistry is gained by performing these fun and safe experiments, the fear of experimenting will be replaced with a desire to know more about the subject.

All of the experiments in this book are basic enough for a person not familiar with scientific terms to understand. The results are dramatic but expected, thus no frightening experiences. Mysterious experiments will be performed—clear liquids change into a green blob, pennies acquire green coats, colors disappear, and many others. Not all of the experiments are magical in nature, but all stir the interest of young and old alike with the wonders and the fun of chemistry. It is hoped that these pleasing chemistry experiences will be used to encourage new chemistry students to seek further knowledge not only in the chemistry field, but in science in general.

Even with our vast knowledge of chemistry there is still so very much to be learned and discovered. Few clues are available as to how a simple plant uses water from the soil, carbon dioxide from the air, and light energy from the sun to produce stored food, a process called photosynthesis. There are many opportunities in the field of chemistry for a person with an inquisitive mind and a spirit of adventure. Much is yet to be learned, but great fun and excitement is in store for the beginning scientist in discovering the chemistry secrets that have already been unlocked.

This book will bring the image of chemistry out of the professional chemistry laboratory and into daily-life experiences. It is designed to present technical chemistry theories in such a way that a person with little or no science training can understand. The experiments are selected based on their ability to be explained basically and on their lack of complexity. A big factor in choosing them was their dramatic appeal. One of the main objectives of the book is to present the *fun* of chemistry.

This is a book for children who want

- *to learn more about chemistry.*
- *to have a collection of fun chemistry experiments.*

The reader will be rewarded with successful experiments if he or she reads an experiment carefully, follows each step in order, and does not substitute equipment. It is suggested that the experiments within a group be performed in order. There is some build up of information from the first to the last, but any terms defined in a previous experiment can be found in the glossary. The following list gives the standard pattern for each exercise:

■ *Purpose: This states the basic goals for the experiment.*

■ *Materials: A list of necessary supplies.*

■ *Procedure: Step-by-step instructions on how to perform the experiment.*

■ *Results: An explanation exactly stating what is expected to happen. This is an immediate learning tool. If the expected results are achieved the experimenter has an immediate positive reinforcement. A "foul-up" is also quickly recognized, and the need to start over or make corrections is readily apparent.*

■ *Why?: An explanation of why the results were achieved is described in understandable terms. This means understandable to the reader who may not be familiar with scientific terms.*

GENERAL INSTRUCTIONS FOR THE READER

■ READ FIRST. *Read each experiment completely before starting.*

■ COLLECT NEEDED SUPPLIES. *Less frustration and more fun will be experienced if all the necessary materials for the experiments are ready for instant use. You lose your train of thought when you have to stop and search for supplies.*

■ EXPERIMENT. *Do not rush into this. Follow each step very carefully, never skip steps and do not add your own. Safety is of utmost importance, and by reading any experiment before starting, then following the instructions exactly, you can feel confident that no unexpected results will occur.*

■ OBSERVE. *If your results are not the same as described in the experiment, carefully reread the instructions, and start over from step one.*

1

1. Kerplunk!

Purpose To demonstrate inertia, a property of matter.

Fact *Matter* can be defined as anything that takes up space and has inertia.

Inertia is a resistance to a change in motion or rest.

Materials *index card*
nickel
drinking glass

Procedure

■ *Lay the index card over the mouth of the glass.*
■ *Place the coin on top of the card. Its position is to be centered over the mouth of the glass.*
■ *Snap the card with your finger.*

Results The card quickly moves forward and the coin drops into the glass.

Why? The stationary card and coin are said to be at rest. They remain motionless because of their inertia. *Inertia* is the tendency of a material not to change its motion or state of rest. When the card is snapped, it slips under the stationary coin. *Gravity* pulls the coin down into the glass.

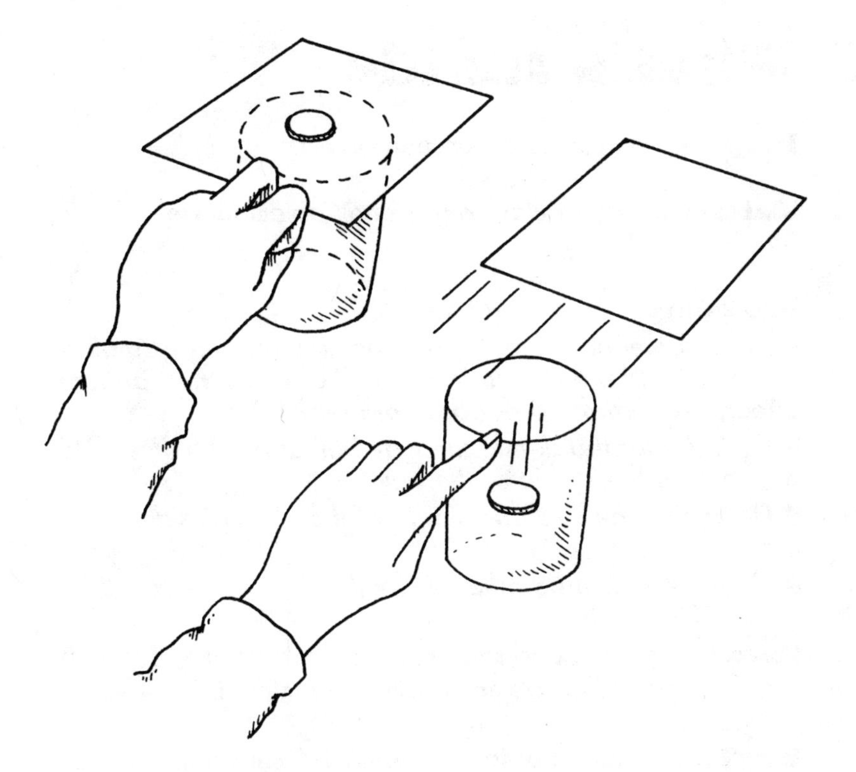

2. What's Inside

Purpose To identify an unseen object.

Materials *ball of clay with a small object inside*
toothpick

Procedure

- *Have someone secretly wrap the clay around a small object and mold the clay into a ball. The object must be firm enough so that the pick does not break it.*
- *Poke the toothpick into the clay ball about 15 times. Do not change the shape of the ball.*
- *Determine the size and shape of the object inside the clay.*
- *Guess what is inside the clay ball.*

Results The size and shape can be determined, and, if it is a familiar object, it will be identified.

Why? The depth that the toothpick is inserted into the clay gives clues to the size and shape of the object. The firmness of the unseen object is determined when the toothpick touches it. Scientists often make decisions about the size and shape of objects without seeing them. You are using a scientific method called *deductive reasoning* to identify the unseen object.

3. Super Chain

Purpose To observe physical properties and their changes.

Facts Physical properties are descriptions about a substance that can be made by seeing, hearing, tasting, feeling, or smelling the material.

Materials *3-×-5-inch index card*
scissors

Procedure

■ *Observe the following physical properties of the index card: color, shape, size, and texture (how it feels).*

■ *Fold the index card in half along the long side.*

■ *Before starting the cuts, study the diagram for the following things:*

a. Notice that all of the cuts are to be about one-quarter inch apart and end at least one-quarter inch from the edge.
b. Notice that the cuts alternate from the folded edge to the open edge.

■ *Start at one end. Make the first cut across the fold stopping one-quarter inch from the open edge.*

■ *The second cut starts at the open edge and stops one-quarter inch from the folded edge.*

■ *Alternate the cuts from the folded side to the open edge. BE CAREFUL TO STOP WITHIN ONE-QUARTER INCH OF EACH EDGE*

■ *Slip the point of your scissors under the fold at point (A) and cut the fold until you reach point (B).* Important: *Do not cut the fold on the two end pieces.*

■ *Carefully stretch the card open to form a large chain.*

■ *Observe these physical properties of the index card again: color, shape, size, and texture.*

Results The color and texture of the card has not changed, but the size and shape of the card has. It was a 3-×-5-inch rectangular, paper card, but after the cutting it resulted in a zig-zag chain-like structure big enough to slip around a person's body.

Why? The procedure for cutting produced the same effect as cutting thin strips from the card and connecting them together. The zig-zag structure allows the paper to stretch out into a large super chain.

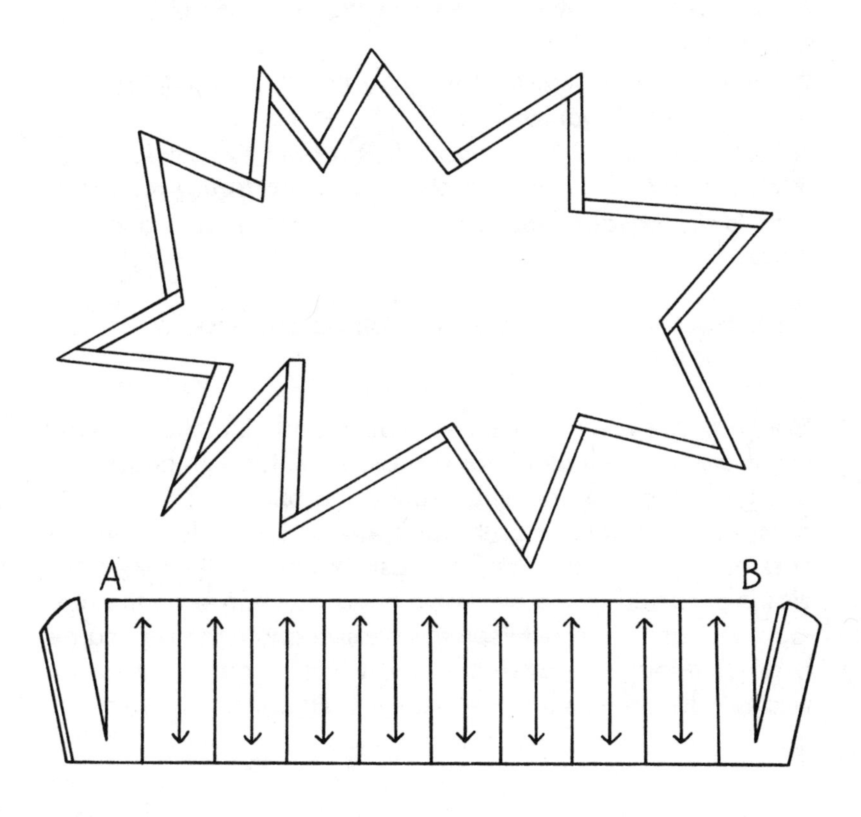

4. Paper Hop

Purpose To illustrate that atoms have positive and negative parts.

Materials *piece of notebook paper*
paper hole punch
balloon (use a size easily held in your hand)

Procedure

■ *Use the hole punch to cut 15 to 20 small circles from the piece of paper.*

■ *Separate the circles and spread them on a table.*

■ *Inflate the balloon and tie it.*

■ *Rub the balloon against your hair, about five strokes. It is important that your hair be clean, dry, and free from oil.*

■ *Hold the balloon close to, but not touching, the paper circles.*

Results The paper circles will hop up and stick to the balloon.

Why? Paper is an example of matter, and all matter is made up of *atoms.* Each atom has a positive center with negatively charged electrons spinning around the outside. The balloon rubs the electrons off of the hair, giving the balloon an excess of negative charges. The positive part of the paper circles is attracted to the excessive negative charge on the balloon. This attraction between the positive and negative charge is great enough to overcome the force of gravity, and the circles will hop upward toward the balloon.

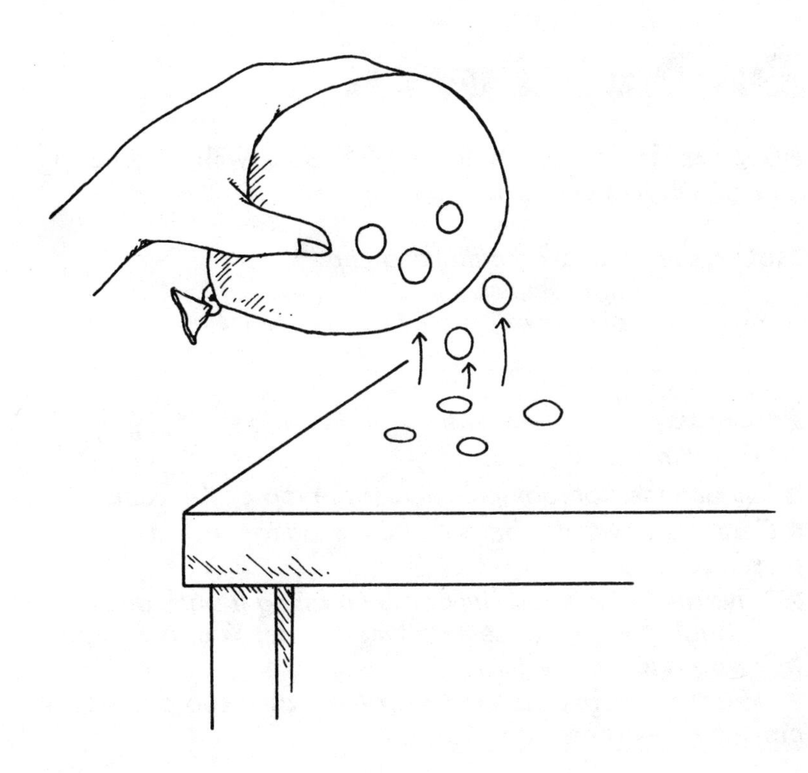

5. Do Not Touch

Purpose To move a balanced toothpick without touching it or any object connected to it.

Materials *clear plastic drinking cup*
flat toothpick
coin (nickel)
balloon

Procedure

■ *Stand the coin up on its edge.*

■ *Balance the flat toothpick across the top of the coin.*

■ *Carefully cover this balanced combination with a clear plastic cup.*

■ *"Charge" an inflated balloon by rubbing it back and forth against your hair several times.* Note: *Your hair must be clean and free of oil.*

■ *Hold the charged balloon near the plastic cup and slowly move it around the cup.*

Results The toothpick moves.

Why? All matter is made of tiny parts called atoms. Each atom has a positively charged center with negatively charged electrons spinning around the outside. Rubbing a balloon on hair causes it to become negatively charged. This charging occurs because some of the electrons from the hair are rubbed off the hair and onto the balloon.

Very little force is needed to move the balanced toothpick. The attractive force between the negatively charged balloon and the positive centers of the toothpick atoms is strong enough to move the pick.

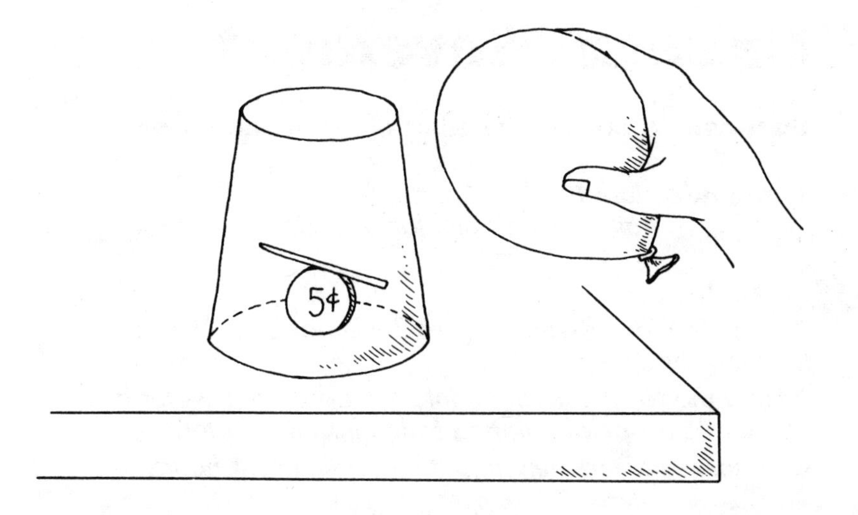
5¢

6. Unseen Movement

Purpose To observe the effect of molecular motion.

Materials *dark food coloring*
tall, one-half pint jar of water

Procedure

■ *Place the jar of water where it will not be moved or touched for 24 hours.*
■ *Hold the food coloring container above the water and allow two drops of coloring to fall into the water.*
■ *Observe immediately and then again in 24 hours.*

Results The drops of coloring sink to the bottom of the jar forming colored streaks in the water as they fall. After 24 hours the water is evenly colored.

Why? The atoms and molecules that make up matter are all in constant motion. Though not seen by the naked eye, water molecules are moving. The small particles of food coloring are being pushed and shoved around by the moving water molecules. Given enough time, the colored particles will be evenly spread throughout the jar of vibrating water. The movement of the color throughout the water is called *diffusion.*

7. An Empty Sack?

Purpose To demonstrate that air is an example of matter and that it takes up space.

Materials *empty plastic bread sack*

Procedure

■ *Fill the empty sack by opening the top and moving the sack through the air.*
■ *Close the top by twisting the opening and holding it with your hand.*
■ *Squeeze the sack with your other hand.*

Results The sack resists being squeezed.

Why? The air molecules fill the sack and apply pressure to the inside. These gas molecules are pushing out more than you are pushing in. If enough pressure could be applied, the molecules would move closer together and the sack would deflate. You will not be able to apply enough pressure for this to happen.

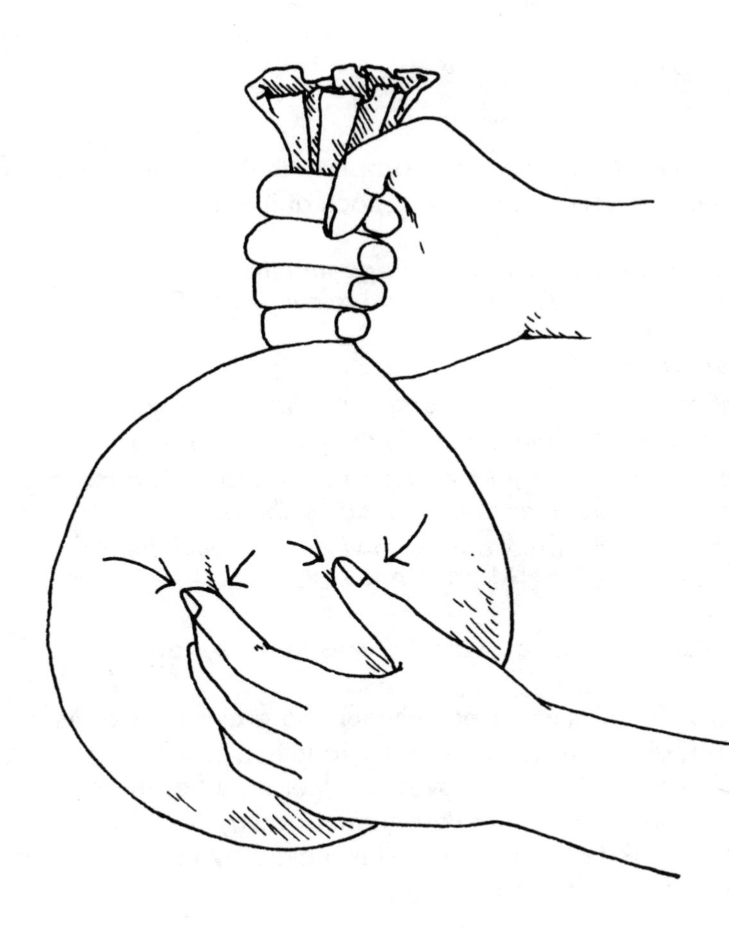

8. A Rising Ball

Purpose To observe the property that no two pieces of matter can occupy the same space at the same time.

Materials *glass quart jar (large-mouthed with a lid)*
small jacks ball or walnut

Procedure

■ *Fill the quart jar one-quarter full with uncooked rice.*
■ *Put the ball or walnut inside the jar and close the lid.*
■ *Hold the jar upright then turn it over.* Note: *Add more rice if the ball cannot be covered by the rice.*
■ *Shake the jar back and forth vigorously until the ball surfaces.* DO NOT SHAKE UP AND DOWN.

Results The ball or walnut comes to the surface.

Why? There are spaces between the grains of rice. As the jar is shaken the rice gets closer together. This is referred to as *settling.* As the rice moves together it pushes the ball upward. Two pieces of matter cannot occupy the same space at the same time; thus the ball is moved by the packing together of the rice grains.

9. Not at the Same Time

Purpose To observe that two pieces of matter cannot occupy the same space at the same time.

Materials *clear drinking glass, 12 oz.*
6 marbles
masking tape

Procedure

- *Fill one-half of the glass with water.*
- *Use a piece of tape to mark the top of the water level.*
- *Very carefully add the marbles to the water by tilting the glass and allowing one marble at a time to slide down the inside to the bottom.*
- *Set the glass upright and notice the water level.*

Results The water level is higher with the marbles in the glass.

Why? Water and marbles are both examples of matter. Two pieces of matter cannot occupy the same space at the same time. When the marbles are dropped into the jar, the water is pushed out of the way by the marbles. The rise in the water level is equal to the volume of the marbles.

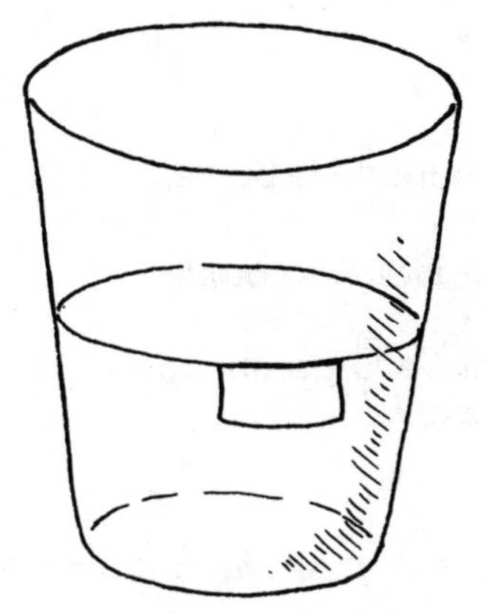

10. No Room

Purpose To try to inflate a balloon inside a bottle.

Materials *cola bottle or any small-mouthed bottle*
balloon
Note: *The balloon must be large enough to fit over the mouth of the bottle.*

Procedure

■ *Hold on to the top of the balloon and push the bottom inside the bottle.*

■ *Stretch the top of the balloon over the mouth of the bottle.*

■ *Try to inflate the balloon by blowing into it.*

Results The balloon only expands slightly.

Why? The bottle is filled with air. Blowing into the balloon causes the air molecules inside the bottle to move closer together, but only slightly. The air is in the way of the balloon, thus preventing it from inflating.

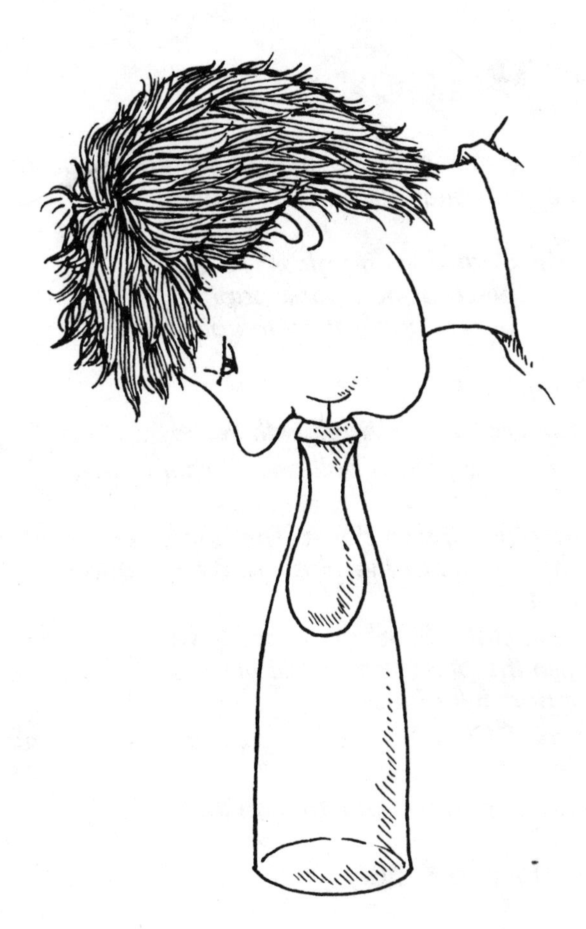

11. Dry Paper

Purpose To demonstrate that even though gases cannot always be seen, they do take up space.

Materials *clear drinking glass, 12 oz.*
piece of notebook paper
bucket (taller than the glass)

Procedure

■ *Fill the bucket one-half full with water.*

■ *Wad the paper into a ball and push it to the bottom of the glass.*

■ *Turn the glass upside down. The paper wad must remain against the bottom of the glass. Make the paper ball a little bigger if it falls.*

■ Important: *Hold the glass vertically with its mouth pointing down. Push the glass straight down into the bucket filled three-quarters full of water.*

■ Important: DO NOT TILT *the glass as you lift it out of the water.*

■ *Remove the paper and examine it.*

Results The paper is dry.

Why? The glass is filled with paper and air. The air prevents the water from entering the glass thus keeping the paper dry.

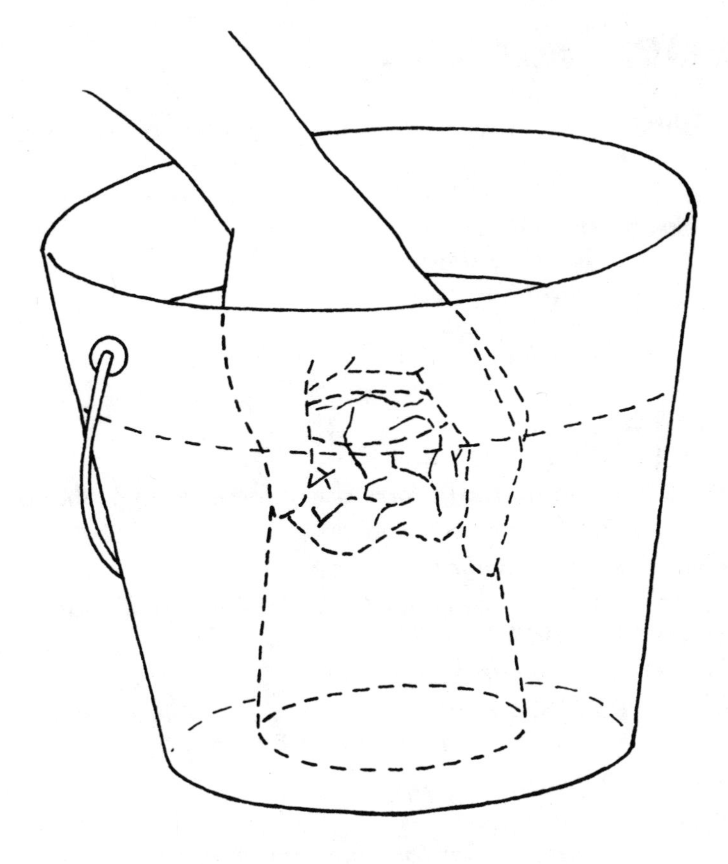

12. How Much?

Purpose To demonstrate that 1 + 1 does not always equal two.

Materials *clear glass quart jar*
1 cup of sugar
measuring cup
masking tape
pencil or pen

Procedure

Making A Measuring Jar

- *Place a strip of masking tape down the outside of the jar.*
- *Pour 1 cup of water into the jar.*
- *Mark the water level on the tape.*
- *Add a second cup of water to the jar and again mark the water level on the tape.*
- *Empty and dry the measuring jar.*
- *Pour 1 cup of sugar into the jar. Make sure that the top of the sugar is at the 1 cup mark on the tape.*
- *Add 1 cup of water.*
- *Stir.*
- *Keep the measuring jar for other experiments.*

Results The liquid level is below the 2 cup mark on the tape.

Why? Water and sugar are examples of matter and cannot occupy the same space at the same time. The cup of sugar is not solid throughout. There are spaces between the sugar grains. The water moves into these spaces resulting in a volume that is less than 2 cups.

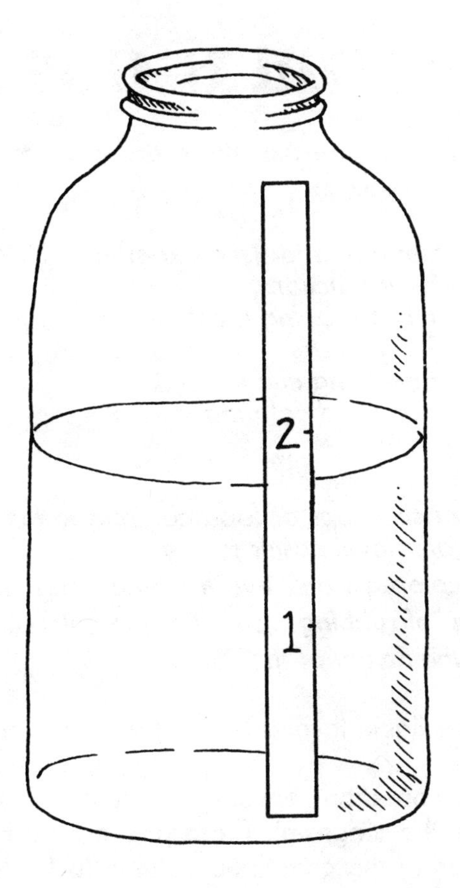
2
1

13. Where Did It Go?

Purpose To illustrate that there are pockets of space between water molecules.

Materials *measuring jar (see experiment* HOW MUCH? *for instructions)*
1 cup rubbing alcohol
1 cup water
measuring cup
blue food coloring

Procedure

- *Add five or six drops of food coloring to the water to make the water level easier to see.*
- *Pour the colored water into the measuring jar.*
- *Add 1 cup of rubbing alcohol to the colored water.*
- *Observe the height of the liquid.*

Results The liquid level is below the 2 cup mark.

Why? The connection of water molecules forms small empty pockets (see the diagram). These pockets are filled with the alcohol, causing the combined volume to be less than two cups.

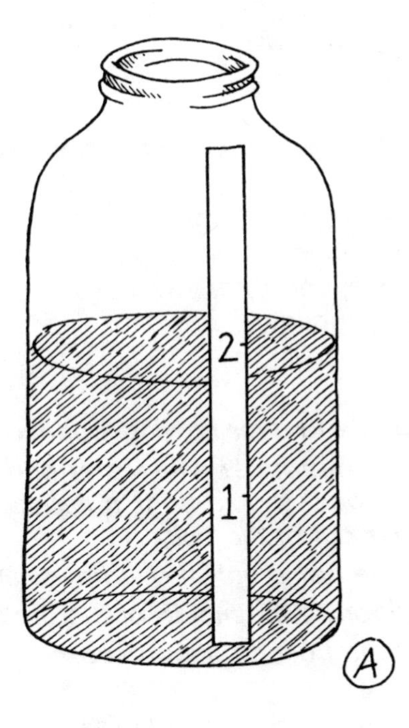
2
1
A

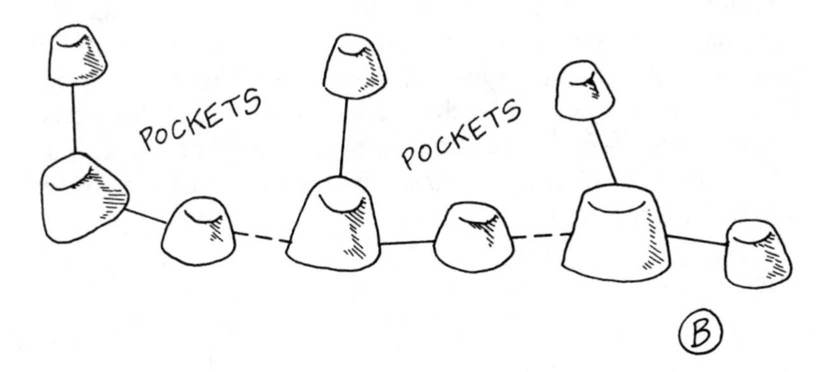
POCKETS
POCKETS
B

14. Sinker

Purpose To sink and raise an eye dropper by changing its density.

Materials *soda bottle*
glass eyedropper
5-inch balloon

Procedure

■ *Fill the soda bottle to over flowing with water.*

■ *Partially fill the eyedropper with water and place it in the bottle. The eyedropper should float, if it sinks, squeeze some of the water out of the bulb.*

■ *Add water to the bottle until it overflows.*

■ *Attach the balloon to the mouth of the bottle.*

■ *Push down on the balloon, then release.*

Results The eyedropper sinks and rises.

Why? Pushing down on the balloon causes water to move into the eyedropper. This extra water makes the dropper heavier and it sinks. Releasing the balloon decreases the pressure in the bottle and the excess water moves out of the dropper and, being lighter, it now rises. The eyedropper changes only its weight by the addition and loss of the water. Since its size remains constant one can say that the density of the dropper changed. Density is a measure of weight for a specific size.

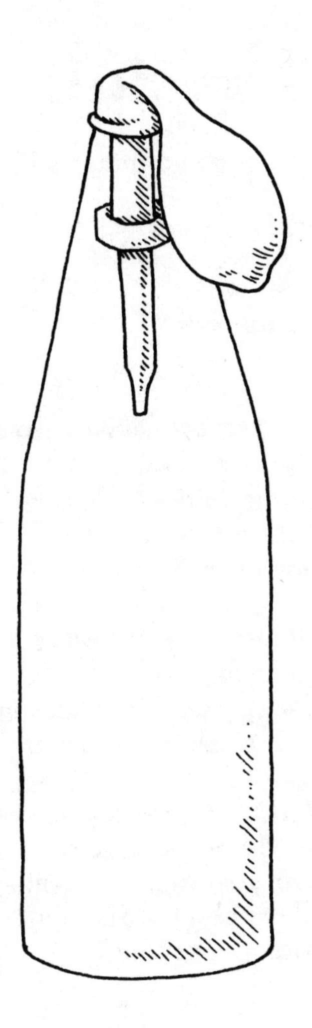

15. Magic Solution

Purpose To float an egg in a magic solution.

Materials *2 clear plastic cups*
3 tablespoons table salt
2 small eggs
¼ teaspoon milk

Procedure

■ *Fill both cups three-quarters full with water.*

■ *Add the milk to one cup of water.*

■ *Add and stir in the salt to the second cup of water. Write the word* MAGIC *on this cup.*

■ *Place an egg in each cup.*

Results The egg floats in the *MAGIC* solution, but sinks in the milky solution.
Note: If the egg does not float in the magic solution, add more salt to the water.

Why? The milk was added only to give the water a cloudy appearance like the *MAGIC* salt water. The egg floats because it is not as heavy as the salty water. The heavy salt water is able to hold the egg up. The egg in the milky water is heavier than the water; thus it sinks.

2

Forces

16. No Heat

Purpose To make water appear to boil with only the touch from a finger.

Materials *cotton handkerchief*
clear drinking glass (with straight, smooth sides)
rubber band

Procedure

■ *Wet the handkerchief with water. Squeeze out any excess water.*

■ *Fill the glass to the top with water.*

■ *Drape the wet cloth over the mouth of the glass.*

■ *Place the rubber band over the cloth in the middle of the glass to hold the cloth close to the glass.*

■ *Use your fingers to push the cloth down about one inch below the water level.*

■ *Pick the glass up, hold the bottom with one hand, and turn it upside down.* Note: *There will be some spillage, so do this over a sink.*

■ *Place the other hand under the hanging cloth and hold the glass. At this point one hand is holding the cloth next to the glass with the free end of the cloth draped over this hand.*

■ *With the free hand push down on the bottom of the glass. Allow the glass to slowly slip down into the cloth.*

Results The water does not fall out of the glass and it appears to start boiling.

Why? Water does not flow out of the cloth because the tiny holes in the cloth are filled with water. Water molecules have a strong attraction for each other which draws them close together. This causes the water to behave as if a thin skin were covering each hole in the cloth, preventing the water in the glass from falling out.

Pushing the glass down causes the cloth to be pulled out of the glass. This outward movement of the cloth creates a vacuum inside and the air outside is pushed through the cloth. Small bubbles of air form inside the water, giving an appearance of boiling water.

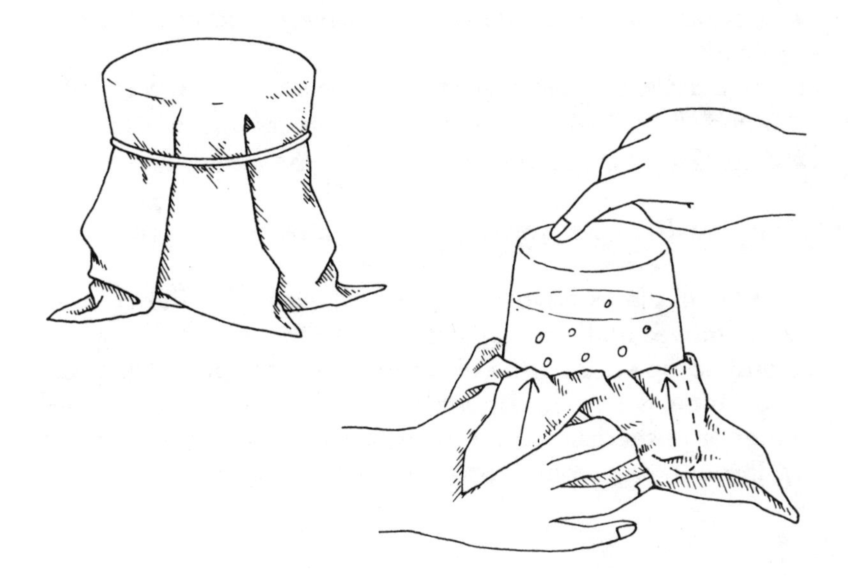

17. Rising Water

Purpose To change the color of celery leaves.

Materials *1 stalk of fresh celery with leaves*
green food coloring
1 clear drinking glass

Procedure

■ *Fill the glass about one-quarter full with water.*

■ *Make a dark green solution by adding food coloring to the water.*

■ *Cut across the bottom end of the celery stalk with a knife.*

■ *Stand the cut end of the celery in the colored water.*

■ *After 24 hours observe the color of the leaves.*

Results The pale green leaves are now a dark green.

Why? All plants have tiny tubes in their stalks. The colored water moves up through these tubes to the leaves. The water is pushed upward by the air pressure in the room. The pressure inside is less than outside the tubes; thus, the colored water is pushed up to the leaves. The movement of water up through tiny tubes is called *capillary action.*

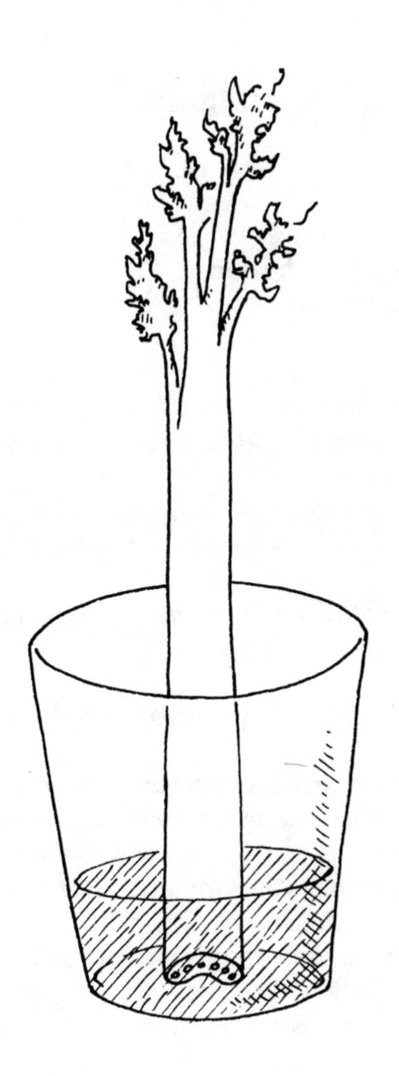

18. **Floating Sticks**

Purpose To observe the pulling power of water molecules.

Materials *3 toothpicks*
liquid dish soap
quart glass bowl

Procedure

■ *Fill the bowl three-quarters full with water.*

■ *Place two toothpicks side by side on the surface in the center of the water.*

■ *Treat the third toothpick by dipping its point in liquid detergent.* Note: *Only a very small amount of detergent is needed.*

■ *Touch the treated toothpick tip between the floating sticks.*

Results The sticks *quickly* move away from each other.

Why? The surface of water acts as if a thin skin were stretched across it. This allows objects to float on top. Detergent breaks the attraction between the molecules where it touches, causing the water molecules to move outward and taking the floating sticks with them. This outward movement occurs because the water molecules are pulling on each other. It is almost as if the molecules are all playing tug of war, and any break causes the "tuggers" to fall backwards.

Dianne
DETERGENT

19. Tug of War

Purpose To demonstrate the difference in the pulling power of water and alcohol.

Materials *1-foot sheet of aluminum foil*
food coloring (red or blue)
rubbing alcohol
water
eyedropper
2 cups

Procedure

■ *Add enough food coloring to ½ cup of water to make a dark solution.*

■ *Fill a second cup one-quarter full with alcohol.*

■ *Smooth the sheet of aluminum foil on a table.*

■ *Pour a very thin layer of the colored water onto the foil.* Note: *The thinner the water, the better.*

■ *Use the eyedropper to add a drop of alcohol to the center of the thin layer of colored water.*

Results The water rushes away from the alcohol leaving a very thin layer of alcohol on the foil. The water is pulling and this causes a pulsation around the edges of the alcohol.

Why? The water molecules on the surface of the water are pulling equally in all directions before the alcohol is added. When the drop of alcohol touches the water there is an immediate separation between the two liquids. Alcohol is pulling away from the water and the water is pulling away from the alcohol. The water molecules seem to be victorious and the water spreads outward taking some of the alcohol with it. This outward movement causes the alcohol to be spread into a thin layer over the foil. It also causes the water molecules to stack up and form a ridge around the alcohol layer. This ridge has a pulsating motion because the water and alcohol molecules continue to pull on each other. The pulling stops when the two liquids totally mix together.

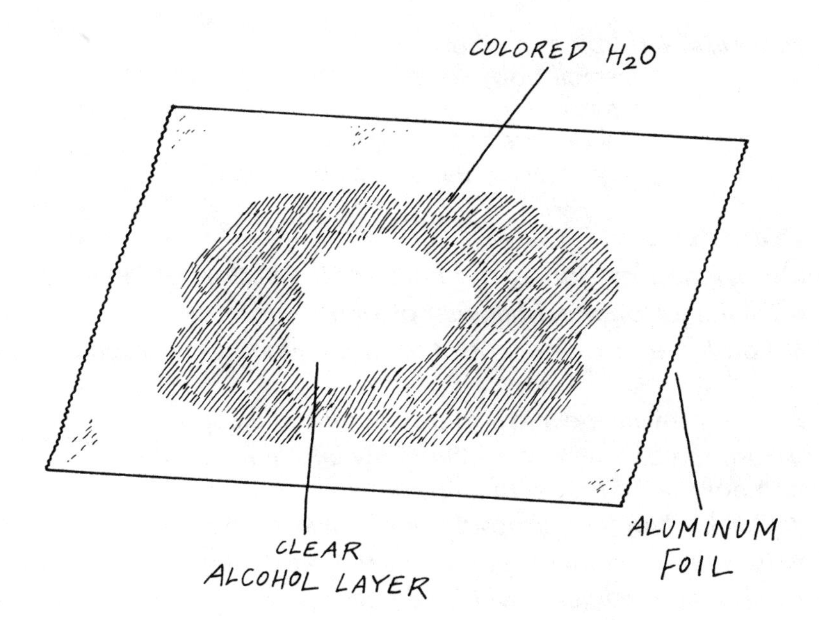

20. Gravity Won

Purpose To demonstrate the effect of gravity on weak surface tension.

Materials *rubbing alcohol*
1 small baby food jar
1 straw
red or blue food coloring
clay, a piece the size of a marble.

Procedure

- *Press the clay against the inside of the bottom of the jar.*
- *Fill the jar one-half full with alcohol.*
- *Add three or four drops of food coloring to the alcohol and stir.*
- *Slowly lower the straw into the colored alcohol.*
- *Push the bottom end of the straw into the clay. The straw can now stand in a vertical position.*
- *Quickly turn the jar upside down over a sink.*
- *Turn the jar right side up and set it on a table.*
- *Observe the liquid level inside the straw.*

Results The colored alcohol flows out of the jar and out of the straw.

Why? The attraction between the alcohol molecules is not very great, and the pressure of the air in the straw is not enough to hold up the liquid. The downward force of gravity pulls the alcohol out of the straw. Perform the experiment in this book called *ANTI-GRAVITY?* to compare the results of using alcohol to those of using water.

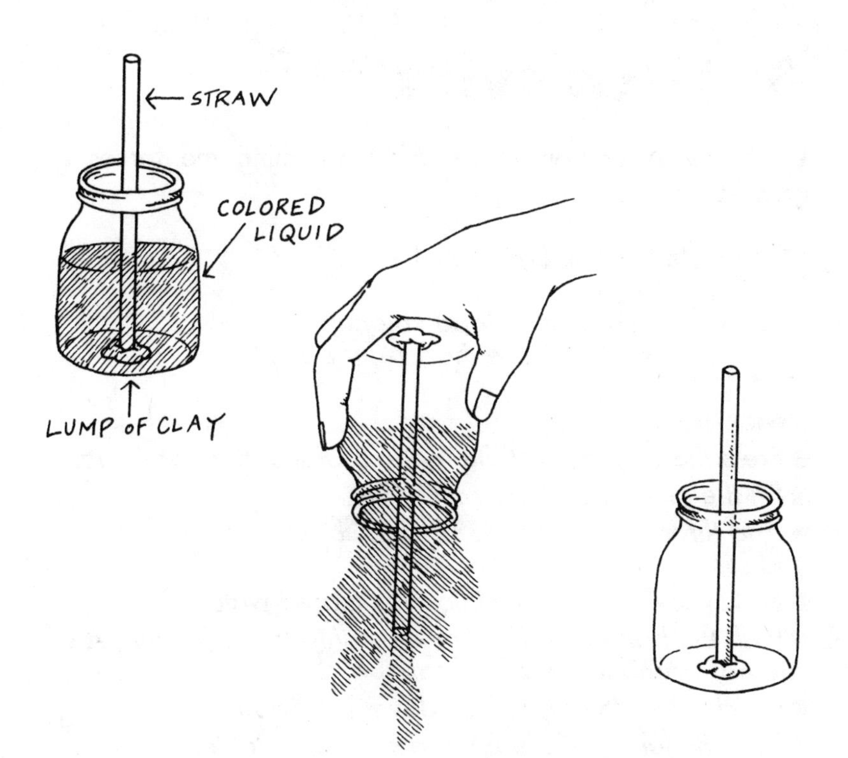
STRAW
COLORED LIQUID
LUMP OF CLAY

21. Anti-Gravity?

Purpose A demonstration of overcoming the forces of gravity.

Materials *1 small baby food jar*
1 straw
red or blue food coloring
clay, a piece the size of a marble

Procedure

- *Press the clay against the inside of the bottom of the jar.*
- *Fill the jar one-half full with water.*
- *Add three or four drops of food coloring to the water and stir.*
- *Slowly lower the straw into the colored water.*
- *Push the bottom end of the straw into the clay. The straw can now stand in a vertical position.*
- *Quickly turn the jar upside down over a sink.*
- *Turn the jar right side up and set it on a table.*
- *Observe the liquid level inside the straw, if any.*

Results Colored water remains in the straw. The height of the water in the straw is the same as that of the water before it was poured out.

Why? Water molecules are attracted to each other. At the surface of the water the molecules tug on each other so much that a skin-like surface is produced. The air in the straw pushes up on the water when the jar is inverted and water molecules are pulling from side to side. These forces are greater than the downward force of gravity; thus the water remains in the straw. Perform the experiment in this book called *GRAVITY WON* to compare the results of using water to that of using alcohol.

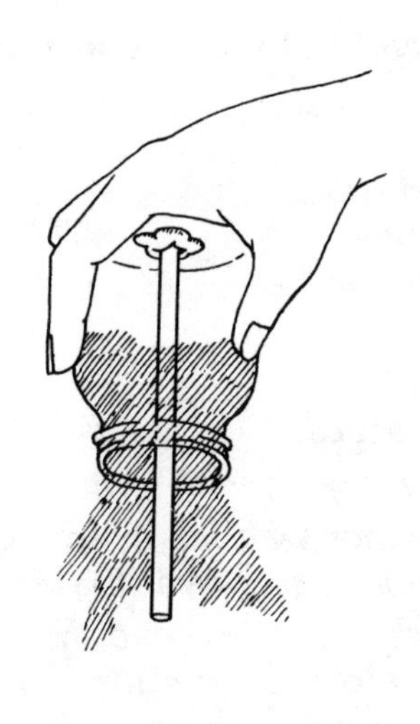

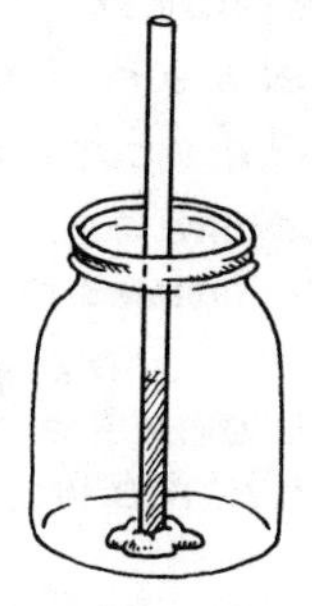

22. Over the Rim

Purpose To observe that water can rise above the edge of its container.

Materials *cup and saucer*
paper clips
eyedropper

Procedure

■ *Set the cup in the saucer.*

■ *Fill the cup to overflowing with water.*

■ *Continue to add water with the eyedropper until one drop causes the water to spill over the rim.*

■ *Drop one paper clip at a time into the cup until the water spills over the edge.* Note: *Before the water spills over, look at the water's surface from the side.*

Results The water rises above the rim of the cup. The height of the water continues to rise as the paper clips are added. The water finally spills over the rim.

Why? Water molecules across the surface are attracted to each other. This attraction is strong enough to allow the water to rise above the top of the cup without spilling. The bulge of water above the rim finally gets so high that the molecules of water can no longer hold together, and over the rim they go.

PAPER CLIPS

23. Mind of Its Own

Purpose To observe the movement of paper circles that seem to have a mind of their own.

Materials *piece of notebook paper*
paper hole punch
small glass with no more than a 2-inch diameter (candle holder or an egg holder will work)
eyedropper
toothpick

Procedure

■ *Use the hole punch to cut three or four circles from the paper.*

■ *Fill the glass about three-quarters full with water.*

■ *When the water is calm, place paper circle on the surface in the center.*

Results After a few seconds the paper moves to the side.

■ *Add two more paper circles and using the toothpick, move the circles to the center of the water.*

Results The paper continues to move toward the edge.

■ *Remove the paper and fill the glass to overflowing with water. Use the eyedropper to add the extra drops needed to make the water bulge above the sides of the glass.*

■ *When the water is calm, place the paper circles in the center.*

■ *Use the toothpick to move the circles toward the edge carefully; then release them. Be sure that you do not force the water over the edge of the glass. Repeat.*

Results The paper continues to move toward the center of the water.

Why? Surface water molecules pull on each other, but they are more attracted to the molecules in the glass. This attraction causes the water to be pulled toward the glass. The water on the paper that is placed in the partially filled glass is pulled toward the edge, carrying the lightweight paper circle with it. The glass that is overfilled with water does not have the exposed glass sides for the water to be attracted to. The result is that the water molecules pull on each other with the force directed toward the center of the water's bulge. The wet paper is pulled toward the center because the water on it is pulled in this direction.

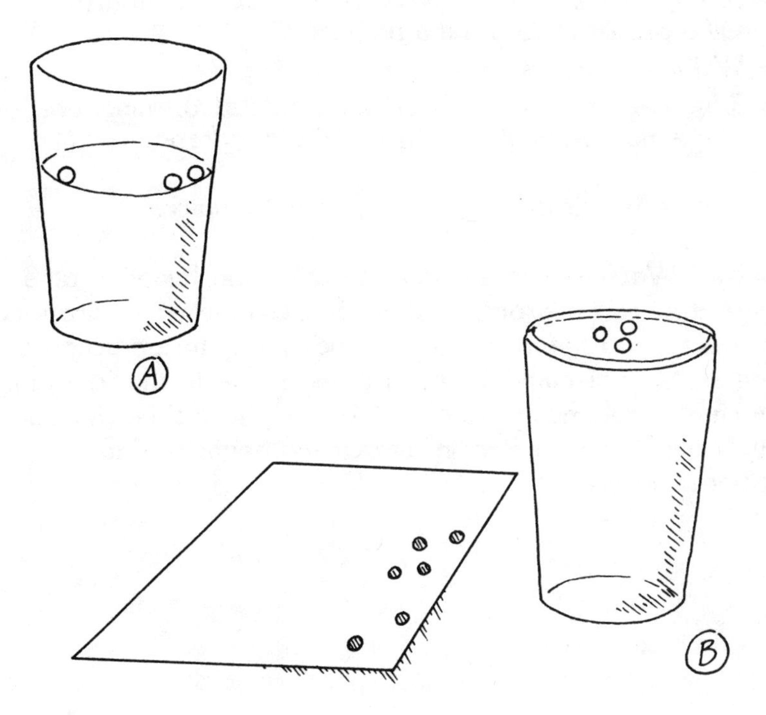

24. Moving Drop

Purpose To demonstrate the attractive force between water molecules.

Materials *1-foot sheet of wax paper*
toothpick
eyedropper
water

Procedure

■ *Spread the wax paper on a table.*

■ *Use the eyedropper to position three or four separate small drops of water on the paper.*

■ *Wet the toothpick with water.*

■ *Bring the tip of the wet pick near, but not touching, one of the water drops. Repeat with the other drops.*

Results The drop moves toward the toothpick.

Why? Water molecules have an attraction for each other. This attraction is strong enough to cause the water drop to move toward the water on the toothpick. The attraction of the water molecules for each other is due to the fact that each molecule has a positive and a negative side. The positive side of one molecule attracts the negative side of another molecule.

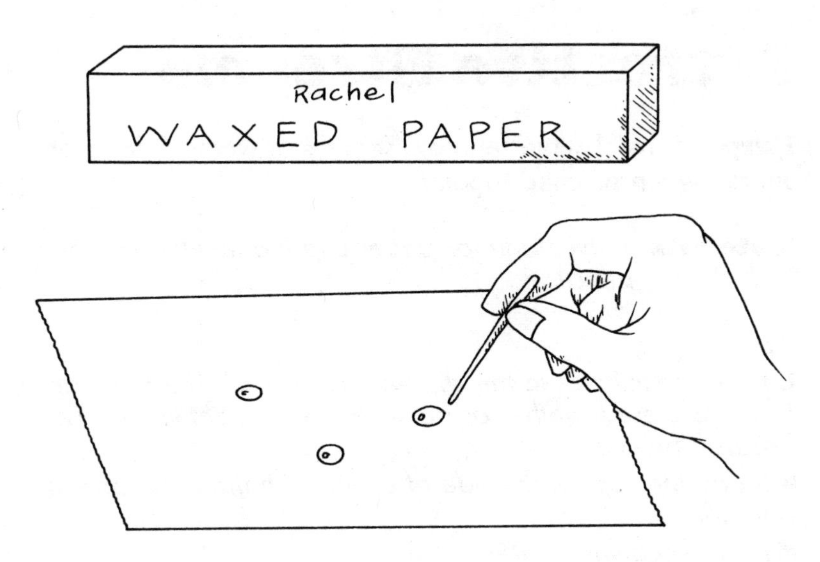
Rachel
WAXED PAPER

25. Attractive Streams

Purpose To observe separate streams of water forming one stream when pinched together.

Materials *1 styrofoam or paper cup (no less than 6 oz.)*
pencil

Procedure

■ *Punch four holes in the cup with the pencil. The holes are to be as close together as possible in a straight line at the base of the cup.*

■ *Stand the cup on the side of a sink with the holes on the sink side.*

■ *Fill the cup with water.*

■ *Take your thumb and forefinger and pinch the four streams of water together.*

Results Water pours out of the four separate holes. Pinching the streams causes the streams to unite. If the holes are close enough, one stream forms. If not, two streams will be formed.

Why? One water molecule has an attraction for another water molecule. These molecules will actually pull on each other and stick together.

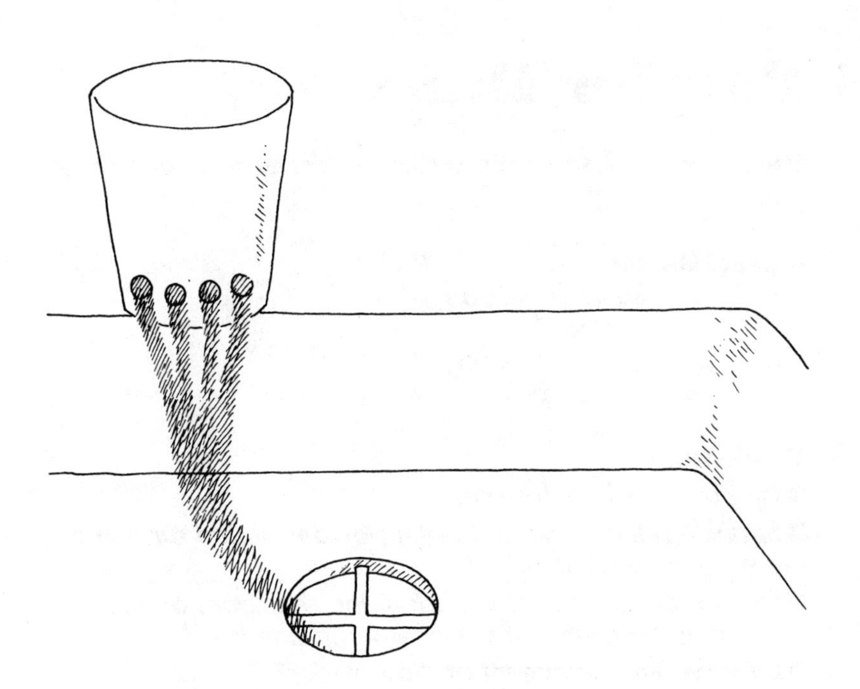

26. Powder Dunk

Purpose To observe the wetting effect of soap and detergent.

Materials *shampoo*
liquid dish soap
toothpicks
talcum powder
2 soup bowls

Procedure

- *Fill both bowls with water.*
- *Sprinkle a thin layer of talcum powder on the surface of the water in each bowl.*
- *Dip the end of one toothpick in the shampoo, and touch the end in the center of the powder in one bowl.*
- *Observe the movement of the powder.*
- *Dip the end of a second toothpick in the liquid dish soap and touch the end in the center of the powder in the second bowl.*
- *Observe the movement of the powder.*

Results Shampoo makes the talcum powder break like large floating ice blocks. The powder rushes to the sides of the bowl and starts to sink when touched by liquid dish soap.

Why? Talcum powder is water resistant. The grains of powder float on top of the water. The water molecules on the surface are pulling equally in all directions before the shampoo or dish soap is added. The addition of the shampoo or dish soap breaks the attraction between the water molecules wherever it touches, causing the water to move outward and taking the floating powder with it. The shampoo is a pure soap while the liquid dish soap is actually a detergent. Detergents are wetting agents and soap is not. A wetting agent is able to spread rapidly over the surface of a solid and penetrate the surface. The liquid dish soap dissolves in the water and quickly covers the grains of talcum powder, causing them to sink to the bottom of the bowl.

27. Magic Paper

Purpose To observe the attraction between molecules.

Materials *rubber cement*
1 sheet of newspaper
scissors (These must be strong and sharp. DO NOT *use school scissors.)*
talcum powder

Procedure

■ *Lay the newspaper on a table.*

■ *Evenly spread a thin, but solid covering of rubber cement over one-half of the newspaper page.* Note: *It is important not to leave spaces uncoated or globs of glue in places.*

■ *Allow the rubber cement to dry for five minutes. It will feel tacky.*

■ *Sprinkle talcum powder evenly over the tacky cement.*

■ *Gently rub the powder to make sure that all of the cement is covered.*

■ *Cut the powdered section into one inch strips.*

■ *Hold two of the strips with the powdered sides touching.*

■ *Cut across the ends of the papers.* Important: *Do not try to snip the paper with the ends of the scissors. Insert the paper as far into the scissors as possible, cutting with the largest part of the blade.*

■ *Gently raise the end of one of the strips.*

■ *Hold only the raised edge, allowing the strip to hang.*

Results Instead of two separate short strips there is one long one.

Procedure

■ *Hold the strips with the powdered sides together again.*

■ *Use the scissors to make a 45-degree diagonal cut across the ends of the paper strips.*

■ *Gently raise the end of one strip.*
■ *Hold only the raised edge, allowing the strip to hang.*

Results The paper strips are connected at a 45-degree angle to each other.

Why? The powder is used to cover the cement so that the pieces do not stick together. The sharp edges of the scissors cut the paper. The pressure applied by the blades pushes a small amount of rubber cement along the cut surface. The molecules of the cement have a strong attraction for each other. These molecules are able to bridge the gap between the cut pieces and hold them together.

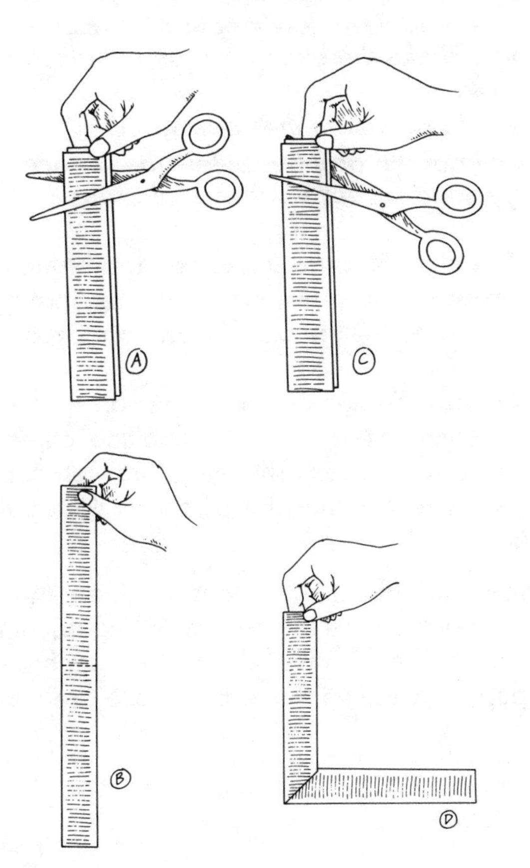

28. Spheres of Oil

Purpose To demonstrate that gravity has little effect on bodies submerged in a liquid.

Materials *clear drinking glass*
½ cup rubbing alcohol
½ cup water
liquid cooking oil
eyedropper

Procedure

■ *Pour ½ cup of water into the glass.*

■ *Tilt the glass and very slowly pour in ½ cup alcohol. Be careful not to shake the glass because the alcohol and water will mix.*

■ *Fill the eyedropper with the cooking oil.*

■ *Place the tip of the dropper below the surface of the top alcohol layer and squeeze out several drops of oil.*

Results The alcohol forms a layer on top of the water. The drops of oil form perfect spheres that float in the center below the alcohol and on top of the water.

Why? Alcohol is lighter and will float on the water if the two are combined very carefully. Shaking causes them to mix, forming one solution. The oil is heavier than alcohol, but lighter than water; thus the oil drops float between the two liquids.

Gravity does not affect the drops because they are surrounded by liquid molecules that are pulling equally on them in all directions. The oil molecules pull on each other, forming a shape that takes up the least surface area, a sphere.

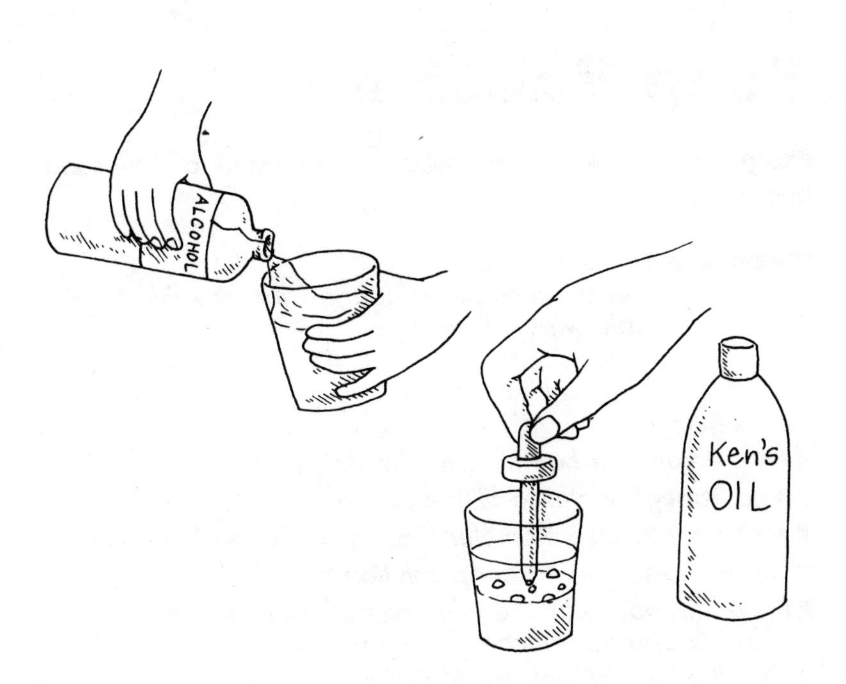
ALCOHOL
Ken's
OIL

29. Soap Bubbles

Purpose To make a soap bubble solution and to blow soap bubbles.

Materials *liquid dish soap*
9-inch piece of 20-gauge wire, any thin bendable wire will work.
cup

Procedure

- *Fill the cup one-half full with the dish soap.*
- *Add enough water to fill the cup. Stir.*
- *Make a 1½-inch-diameter loop in one end of the wire.*
- *Dip the loop into the soap solution.*
- *Hold the loop, with the thin layer of soap stretched across it, about four inches from your mouth.*
- *Gently blow through the film of soap.*

Results Bubbles of soap should be produced. If the soap film breaks, try blowing more gently. Add 1 tablespoon of soap to the solution if the bubbles continue to break. More soap should be added until bubbles are produced.

Why? The soap and water molecules link together to form a ziz-zag pattern. This irregular pattern allows the thin layer of liquid to stretch outward when blown into.

3

Gases

30. Escaping Bubbles

Purpose To determine why bubbles escape from a glass of soda

Materials *small baby food jar*
soda, any flavor carbonated beverage

Procedure

- *Fill the jar one-half full with soda*
- *Set the jar on a table and observe the liquid.*

Results Small bubbles of gas continuously rise to the top of the liquid.

Why? Carbonated beverages are made by dissolving large amounts of CO_2 in flavored water. This excess amount of CO_2 is able to stay in the liquid because it is pushed with high pressure into the bottle and the bottle is immediately sealed. The bubbles that are rising in the cola are escaping CO_2.

SODA

31. Foamy Soda

Purpose To observe gas bubbles being pushed out of a soda by particles of salt.

Materials *small baby-food jar*
1 teaspoon table salt
soda, any flavor carbonated beverage

Procedure

■ *Fill the jar one-half full with the soda.*
■ *Add 1 teaspoon of salt to the soda.*

Results Bubbles form in the liquid, then foam appears on top of the soda.

Why? Each bubble seen in the soda is a collection of carbon dioxide gas. Salt and carbon dioxide are both examples of matter and matter takes up space. When the salt is added to the cola, it pushes the bubbles of carbon dioxide out of its way. These bubbles rise to the top bringing small amounts of soda with them. This movement of the gas forms the foam on top of the liquid. Replacing a gas with another substance is called effervescence.

Bobo
SALT
SODA

32. Pop Cork

Purpose To shoot a cork from a soda bottle.

Materials *soda bottle*
petroleum jelly
½ package dry yeast
1 teaspoon sugar
cork that fits the soda bottle

Procedure

- *Pour ½ package of yeast into the soda bottle.*
- *Fill the bottle one-half full with warm water.*
- *Add 1 teaspoon of sugar.*
- *Place your thumb over the bottle's mouth and shake the bottle vigorously to mix the contents.*
- *Cover the sides of the cork with petroleum jelly.*
- *Loosely stopper the bottle with the cork.*
- *Place the bottle on the ground.*

Results After a few minutes the cork pops out of the bottle and into the air.

Why? Yeast contains tiny plants that use sugar and oxygen to produce energy. As this energy is produced carbon dioxide is also formed. As the amount of carbon dioxide gas increases inside the closed bottle, the pressure of the gas builds. When enough gas is formed, the cork will be pushed out with enough force to produce a popping noise.

SO
SOD

33. Limewater

Purpose To make a testing solution for carbon dioxide.

Materials *lime (used in making pickles)*
tablespoon
2 glass quart jars with lids

Procedure

- *Fill one jar with water.*
- *Add 1 tablespoon of lime and stir.*
- *Secure the lid; allow the solution to stand overnight.*
- *Decant (pour off) the clear liquid into the second jar. Be careful not to pour any of the lime that has settled on the bottom of the jar.*
- *Keep the jar closed. This limewater will be used in other experiments to test for the presence of carbon dioxide.*

Results The liquid is milky white and opaque at first. Large particles of lime start to precipitate. *Precipitate* means to fall downward. After standing overnight the liquid is very clear.

Why? Opaque means that light cannot pass through, and thus one cannot see through it. The undissolved particles of lime are temporarily suspended in the water, making it appear milky and opaque. It takes time for all of the tiny particles to settle. The clear liquid is a saturated solution of limewater. It must be covered to prevent carbon dioxide in the air from dissolving in it.

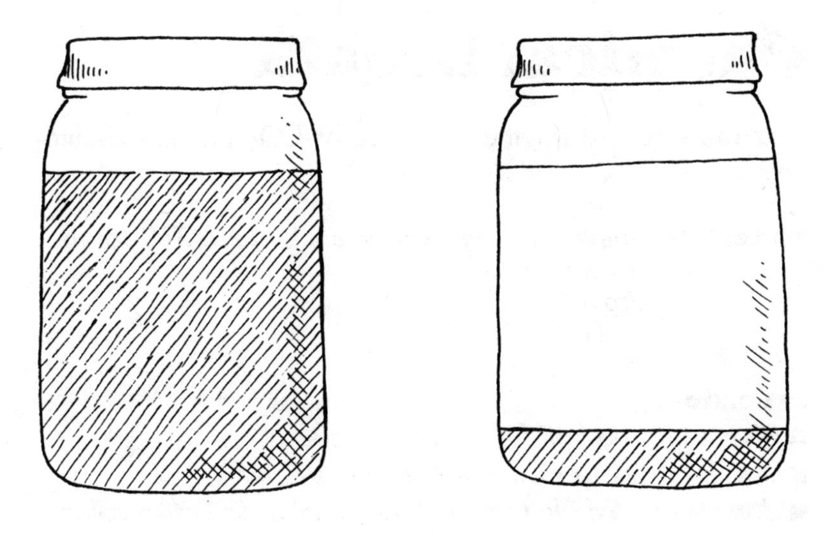

34. Chemical Breath

Purpose To test for the presence of CO_2 gas in exhaled breath.

Materials *limewater (See experiment* LIMEWATER *for instructions.)*
straw
pint jar

Procedure

■ *Fill the jar one-half full with limewater.*

■ *Use the straw to exhale into the limewater.*

■ *Continue to exhale into the liquid until a distinctive color is observed.*

Results The water limewater turns from clear to a milky color.

Why? Limewater always turns milky when CO_2 is mixed with it. The chemical in the limewater combines with the CO_2 gas in the exhaled breath to form a white powder that is not soluble in water. The powder is called limestone. If the solution is allowed to stand for several hours, the powdery limestone will fall to the bottom of the jar.

LIMEWATER

35. A Hungry Plant

Purpose To observe the production of carbon dioxide by yeast.

Materials *soda bottle*
1 teaspoon sugar
½ package powdered yeast
9-inch balloon
18 inches of aquarium tubing
modeling clay
limewater (See experiment LIMEWATER *for instructions.)*

Procedure

■ *Pour ½ package of yeast into the soda bottle.*
■ *Fill the bottle one-half full with warm water.*
■ *Add 1 teaspoon of sugar to the bottle.*
■ *Place your thumb over the bottle's mouth and shake the bottle vigorously to mix the contents.*
■ *Place one end of the aquarium tubing in the top part of the bottle.*
■ *Use the clay to seal off the bottle and to hold the tubing in the bottle.*
■ *Insert the free end of the tube into a glass that is one-half full with limewater.*
■ *Observe periodically for several days.*

Results There is some foaming at first in the soda bottle. Bubbles of gas flow out of the tube into the limewater. The limewater turns cloudy.

Why? Yeast contains tiny plants that use sugar and oxygen to produce energy. In the process of producing this energy carbon dioxide is also formed. The cloudiness of the limewater is proof that the bubbles produced by the reaction is carbon dioxide. Limewater only turns cloudy when carbon dioxide gas is bubbled through it.

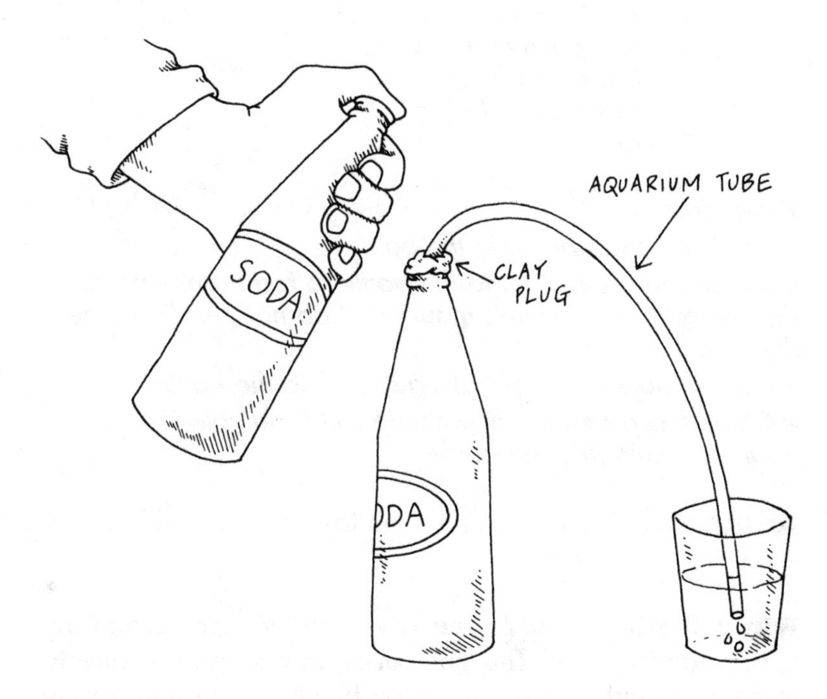

36. Erupting Volcano

Purpose To simulate a volcanic eruption.

Materials *soda bottle*
baking pan
1 cup vinegar
baking soda
red food coloring
dirt

Procedure

■ *Place the soda bottle in the pan.*

■ *Shape moist dirt around the bottle to form a mountain. Do not cover the bottle's mouth and do not get dirt inside the bottle.*

■ *Pour 1 tablespoon of baking soda into the bottle.*

■ *Color 1 cup of vinegar with the red food coloring, and pour the liquid into the bottle.*

Results Red foam sprays out the top and down the mountain of dirt.

Why? The baking soda reacts with the vinegar producing carbon dioxide gas. The gas builds up enough pressure to force the liquid out the top of the bottle. The mixture of the gas and the liquid produces the foam.

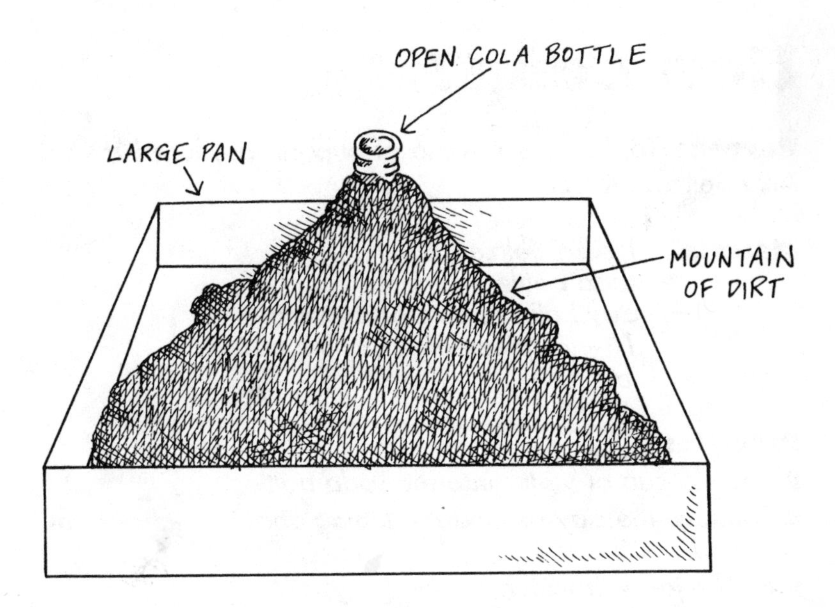
OPEN COLA BOTTLE
LARGE PAN
MOUNTAIN
OF DIRT

37. How Long?

Purpose To time the release of bubbles produced by one Alka-Seltzer tablet.

Materials *1 Alka-Seltzer tablet*
soda bottle
clay ball, the size of a walnut
18-inch piece of aquarium tubing
jar

Procedure

- *Pour 1/4 cup of water into the soda bottle.*
- *Squeeze the clay around the tubing about 2 inches from one end.*
- *Fill the jar with water.*
- *Place the free end of the tube in the jar of water.*
- *Break the Alka-Seltzer tablet into small pieces; quickly drop the pieces into the soda bottle.*
- *Immediately insert the tube into the bottle; seal the opening with the clay.*
- *Record the time.*
- *Watch and record the time when the bubbling stops.*

Results The tablet immediately reacts with the water to produce bubbles. The bubbles are released for about 25 minutes.

Why? The dry acid and baking soda in the tablet are able to combine with the water to form carbon dioxide gas. It is the carbon dioxide gas that moves through the tube and forms bubbles in the glass of water. The bubbling stops when all of the material reacts.

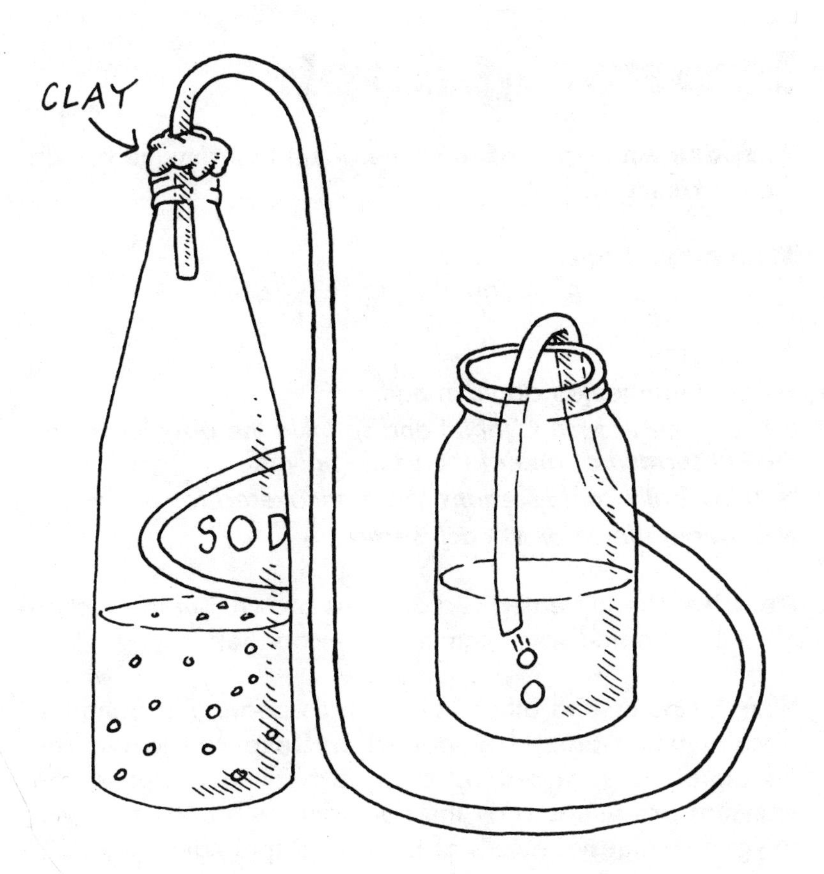
CLAY
SOD

38. Browning Apple

Purpose An investigation of the effect that oxygen has on the darkening of fruit.

Materials *apple*
vitamin C tablet

Procedure

- *Cut the unpeeled apple in half.*
- *Crush the vitamin C tablet and sprinkle the powder over the cut surface of one of the apple halves.*
- *Allow both apple sections to set uncovered for one hour.*
- *Observe the color of each section.*

Results The untreated section turns brown, but the section treated with vitamin C is unchanged.

Why? Apples and other fruit, such as pears and bananas, discolor when bruised or peeled and exposed to air. This discoloration is caused by chemicals called *enzymes.* The enzymes are released by the damaged cells and react with oxygen to digest the cells of the fruit. Rapid color and taste changes occur because of the reaction with oxygen. Vitamin C prevents the darkening by reacting with the enzyme before it can start digesting the cell tissue.

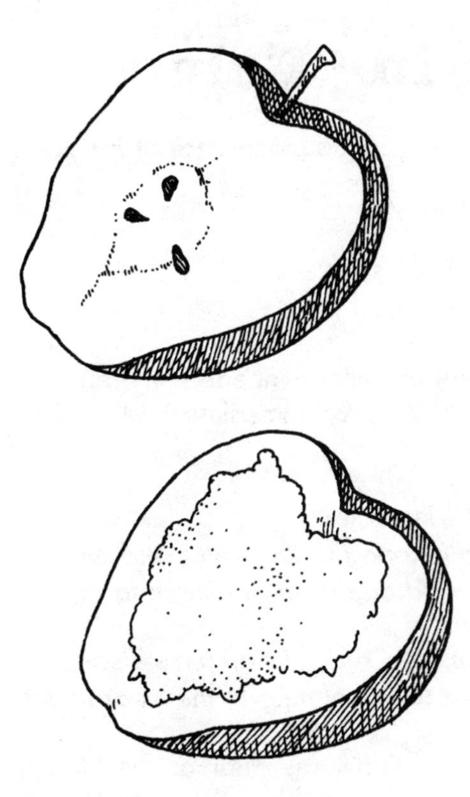

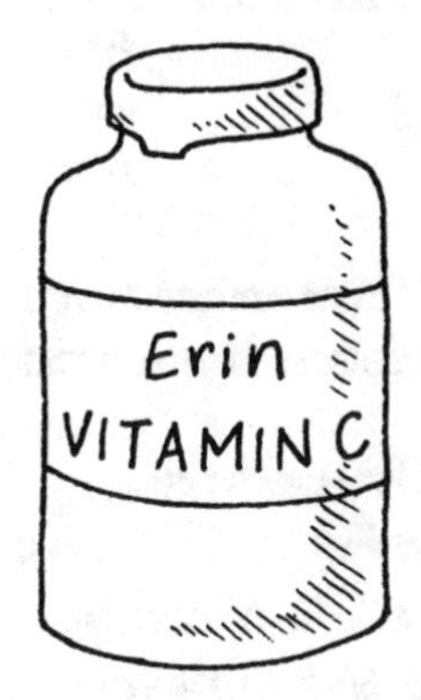
Erin
VITAMIN C

39. Disappearing Color

Purpose To observe the magical disappearance of a color.

Materials *red food coloring*
bleach
eyedropper
small baby food jar

Warning: Adult supervision will be needed in handling bleach. If the bleach spills, clean area immediately with water.

Procedure

- *Fill the jar one-half full with water.*
- *Add two drops of food coloring to the water and stir.*
- *Use the eyedropper to add one drop of bleach to the colored water.*
- *Add drops of bleach until the red solution turns clear.*
- *Now, add a drop of red food coloring to the clear liquid.*

Results The red water solution turns clear as the bleach moves down through it. Adding the red coloring to the clear solution containing bleach produces an interesting magical effect in that the red dye disappears when it hits the liquid.

Why? Bleach contains a chemical called sodium hypochlorite. This chemical contains oxygen that is easily released. Oxygen combines with the chemicals in dyes to form a colorless compound. The bleach decolorizes the red water as it moves downward. The red drop disappears because it is surrounded by bleach which decolorizes the red dye.

BLEACH
RED

40. **Fading Color**

Purpose To observe the effect of dry bleach on color.

Materials *small baby food jar*
red food coloring
powdered bleach
teaspoon

Procedure

- *Fill the jar with water.*
- *Add one drop of food coloring; stir.*
- *Add and stir in one teaspoon of powdered bleach.*
- *Wait for 15 minutes.*

Results The red color starts to fade and finally disappears. The water becomes clear except for any undissolved bleach.

Why? When added to water the powdered bleach slowly gives off oxygen. The combination of this oxygen with the red dye causes the color to fade until it becomes colorless.

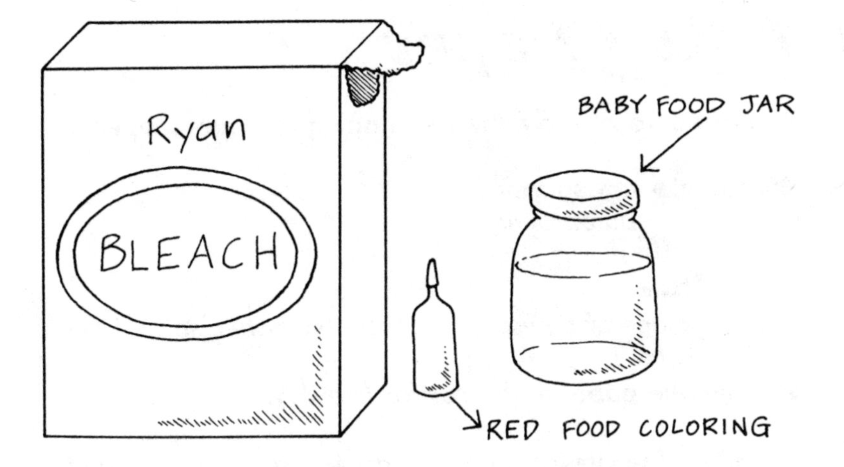
Ryan
BLEACH
BABY FOOD JAR
RED FOOD COLORING

41. Aging Paper

Purpose To observe the rapid aging of a newspaper.

Materials *newspaper*
automobile

Procedure

■ *Lay a piece of newspaper in an automobile so that the sun's rays hit it.*
■ *Leave the paper in the car for five days.*

Results The newspaper appears to have rapidly aged. It changes from white to yellow in color.

Why? This reaction is unique because it is backward from most reactions with oxygen. Usually the addition of oxygen causes the color to become lighter. The materials used to make the newspaper are yellow in color. The chemicals added to turn the paper white do so by removing oxygen. Placing the paper in the car allows the sunlight to heat up the air and the paper causing oxygen to combine with the chemicals in the paper. The addition of the oxygen changes the paper back to its original yellow color. All newspaper will turn yellow after a period of time. The sun's light just speeded up the aging process.

42. Rust Prevention

Purpose To observe the effect that protective coatings have on the rusting of steel wool.

Materials *1 steel wool soap pad*
scissors
plate
1 sheet paper toweling
½ cup vinegar
pencil

Procedure

■ *Cut the steel wool pad into four equal parts.*

■ *Run warm tap water over two of the pieces to remove as much of the soap as possible.*

■ *Place one piece with soap and one piece without soap in the vinegar.*

■ *Mark the paper towel into four equal parts. Number each section.*

■ *Lay the paper towel over a plate.*

■ *Remove the pieces from the vinegar and squeeze out as much liquid as possible.*

■ *Place the steel wool pieces in these indicated sections:*
Section 1: Piece with no soap and soaked in vinegar.
Section 2: Piece with soap and soaked in vinegar.
Section 3: Piece with no soap, but wet with water.
Section 4: Dry piece with soap. This is the control.

■ *Observe the steel wool pieces every 10 minutes for one hour and then allow them to stand for 24 hours.*

Results The piece with no soap that has been soaked in vinegar shows signs of rusting after 10 minutes. It takes up to one hour for the piece with soap that was soaked in vinegar to rust. After 24 hours the vinegar-soaked pieces have equally rusted and the piece wet with water and containing no soap shows only slight rusting. No change is seen in the control. *Note:* A control is any material that is not changed at the start of the experiment.

Why? Steel wool contains iron which rusts by combining with oxygen in the air. Soap helps to prevent air from touching the iron. The vinegar cleans off any additional coating on the steel wool, allowing the iron and oxygen to combine. The iron oxide that is formed is reddish brown in color. One usually thinks of rust as being this color, but other colors are formed when different metals rust by combining with oxygen.

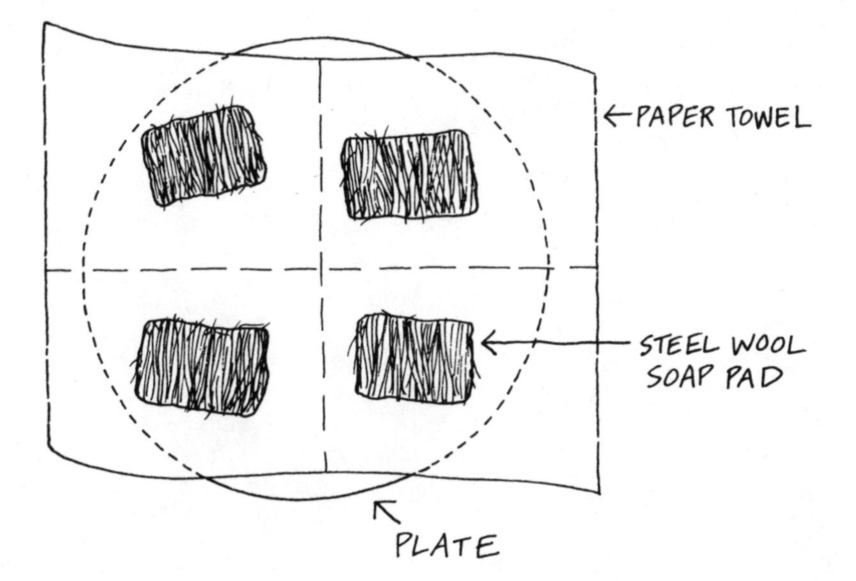

4

Changes

43. Green Pennies

Purpose To give pennies a green coating.

Materials *saucer*
paper towel section
vinegar
3–5 pennies

Procedure

- *Fold the paper towel in half; fold again to form a square.*
- *Place the folded towel in the saucer.*
- *Pour enough vinegar into the saucer to wet the towel.*
- *Place the pennies on top of the wet paper towel.*
- *Wait 24 hours.*

Results The tops of the pennies are green.

Why? Vinegar's chemical name is acetic acid. The acetate part of the acid combines with the copper on the pennies to form the green coating composed of copper acetate.

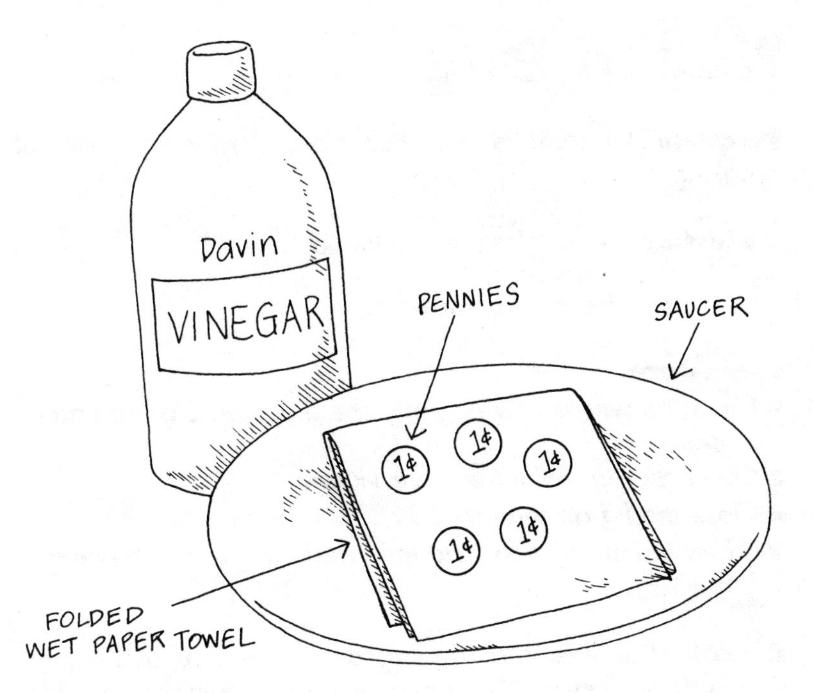
Davin
VINEGAR
PENNIES
SAUCER
1¢
1¢
1¢
1¢
1¢
FOLDED
WET PAPER TOWEL

44. Naked Egg

Purpose To remove the shell from a raw egg without breaking it.

Materials *1-pint glass jar with a lid*
1 raw egg
1 pint of clear vinegar

Procedure

- *Place the whole raw egg into the glass jar. Do not crack the egg.*
- *Cover the egg with the clear vinegar.*
- *Close the lid on the jar.*
- *Observe immediately and then periodically for the next 24 hours.*

Results Bubbles start forming on the surface of the egg shell immediately and increase in number with time. After 24 hours the shell will be gone, and portions of it may be floating on the surface of the vinegar. The egg remains intact because of the thin see-through membrane around the outside. The yolk can be seen through the membrane.

Why? Vinegar's chemical name is acetic acid. Egg shells are made of calcium carbonate. The reaction between acetic acid and calcium carbonate causes the egg shell to disappear and carbon dioxide bubbles to form.

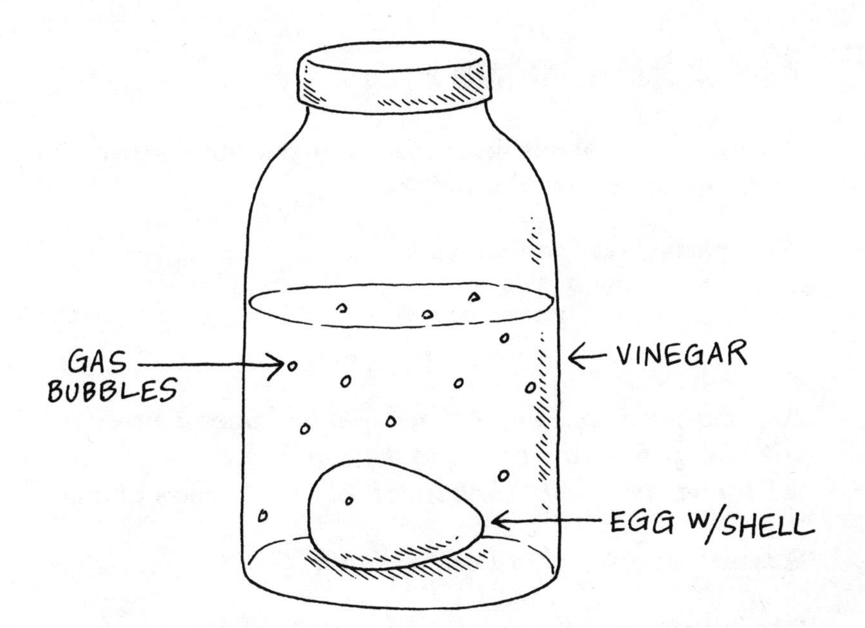
GAS BUBBLES
VINEGAR
EGG W/SHELL

45. Breakdown

Purpose To change hydrogen peroxide into water and oxygen with the aid of a potato.

Materials *hydrogen peroxide*
raw potato
5-oz. paper cup

Procedure

- *Fill the paper cup one-half full with hydrogen peroxide.*
- *Add a slice of raw potato to the cup.*
- *Observe the results. Look specifically for bubbles of gas.*

Results Bubbles of gas are given off.

Why? Raw potatoes contain the enzyme *catalase*. Enzymes are chemicals found in living cells. Their purpose is to speed up the breakdown of complex food chemicals into smaller, simpler, more usable parts. Catalase from the potato's cells causes the hydrogen peroxide to quickly break apart into water and oxygen gas.

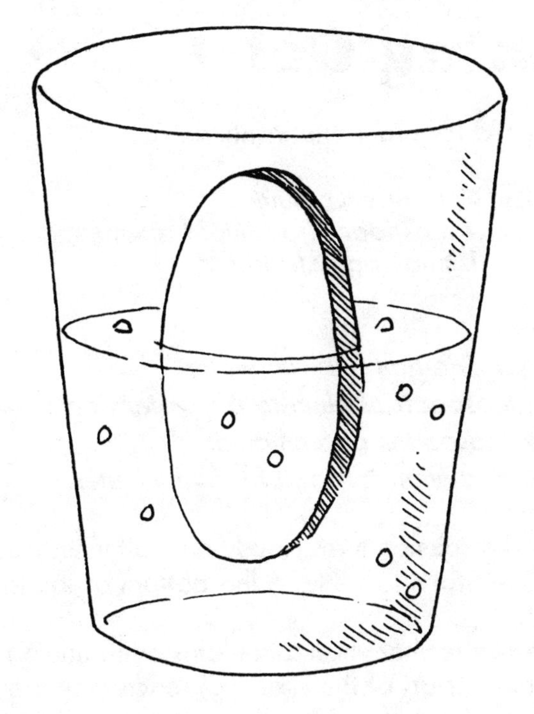

46. Sinking Gel

Purpose To form a white insoluble gel.

Materials *½ teaspoon alum*
2 teaspoons household ammonia
1 small baby food jar

Procedure

- *Fill the jar one-half full with water.*
- *Add ½ teaspoon of alum to the water and stir.*
- *Stir in 2 teaspoons of ammonia.*
- *Allow the solution to stand for five minutes.*

Results The solution turns cloudy and, after standing, a white gel starts to settle to the bottom of the jar.

Why? Household ammonia contains ammonium hydroxide. The hydroxide part of the chemical reacts with the aluminum in the alum. One of the products of the reaction between alum and ammonium is the white, insoluble gel called aluminum hydroxide.

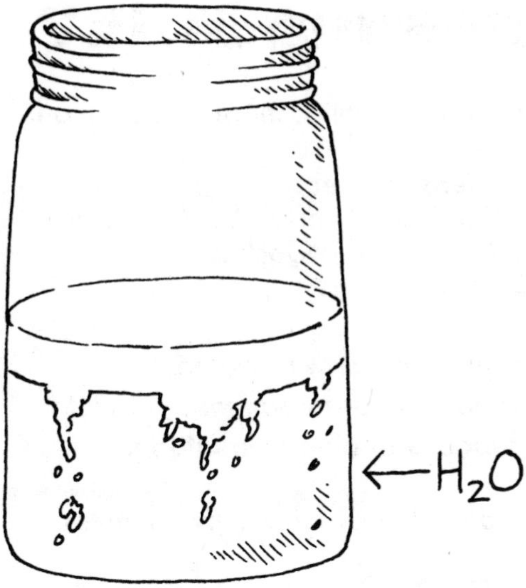

$\leftarrow H_2O$

47. Magnesium Milk?

Purpose To make a milky, magnesia solution.

Materials *1 teaspoon epsom salt*
2 teaspoons household ammonia
1 small baby food jar

Procedure

■ *Fill the jar one-half full with water.*
■ *Stir 1 teaspoon of epsom salt into the water.*
■ *Pour 2 teaspoons of ammonia into the jar.* DO NOT STIR.
■ *Allow the solution to stand for five minutes.*

Results A white, milky substance forms as the ammonia mixes with the epsom salt solution.

Why? Household ammonia's chemical name is ammonium hydroxide. Magnesium sulfate is the chemical name for epsom salt. Mixing ammonia and epsom salt causes a reaction which produces magnesium hydroxide as one of the products. Magnesium hydroxide is a white substance that does not dissolve well in water. After standing awhile, the white floating particles settle to the bottom of the jar. Magnesium hydroxide is part of the medicine called Milk of Magnesia. The name "milk" is used because of the milky appearance.

Ginger
AMMONIA
Jenny
EPSOM
SALT

48. The Green Blob

Purpose To produce a green, jelly-like blob of material from mixing two liquids.

Materials *vinegar*
steel wool
household ammonia
tablespoon
2 small baby food jars with one lid

Procedure

■ *Fill one-half of one jar with steel wool.*

■ *Add enough vinegar to cover the steel wool.*

■ *Secure the lid on the jar; write* IRON ACETATE *on the side of the glass.*

■ *Allow the jar to stand undisturbed for five days.*

■ *Pour one tablespoon of the liquid Iron Acetate into the second jar.*

■ *Add one tablespoon of household ammonia and stir.*

Results A dark green, jelly-like substance forms immediately.

Why? The iron in the steel wool combines with the vinegar to produce iron acetate. Household ammonia's chemical name is ammonium hydroxide. A chemical reaction occurs as soon as these two liquids combine. The word equation for the reaction is:

ammonium hydroxide + iron acetate yields
ammonium acetate + iron hydroxide

Notice that there is an exchange of materials. Nothing new was produced. There is still ammonium, iron, hydroxide, and acetate, but the recombination produces a totally different result. The original materials were liquids and the product is a gel. The starting materials in a chemical reaction break apart and are rearranged to form the products. There are never new basic materials produced.

49. Starch I.D.

Purpose To determine how to test materials for the presence of starch.

Materials *¼ teaspoon flour*
Tincture of Iodine
saucer
tablespoon

Procedure

- *Place ¼ teaspoon of flour in a saucer.*
- *Add 3 tablespoons of water and stir.*
- *Add three or four drops of the Tincture of Iodine.*

Results The combination of starch and iodine produces an intense blue-purple color.

Why? Starch is a very large chemical molecule. It looks like a long twisted chain with many branches sticking out. This long twisted chain is thought to capture the iodine inside its spiral pattern. The spiral of starch with iodine caught on the inside produces the color.

FLOUR
H.L.'s
TINCTURE
OF
IODINE

50. **Testing for Starch**

Purpose To test for the presence of starch in different materials.

Materials *cookie sheet*
eyedropper
Tincture of Iodine
Testing Samples:
notebook paper
cheese
bread
cracker
sugar
apple slice

Procedure
- *Place the testing samples on the cookie sheet.*
- *Place one drop of iodine on each of the testing samples.*

Results The paper, bread, and cracker turn a dark blue-purple. The other samples are just stained by the brown iodine solution.

Why? Starch combines with iodine to form a blue-purple compound. Only the samples containing starch turn dark where the iodine is added.

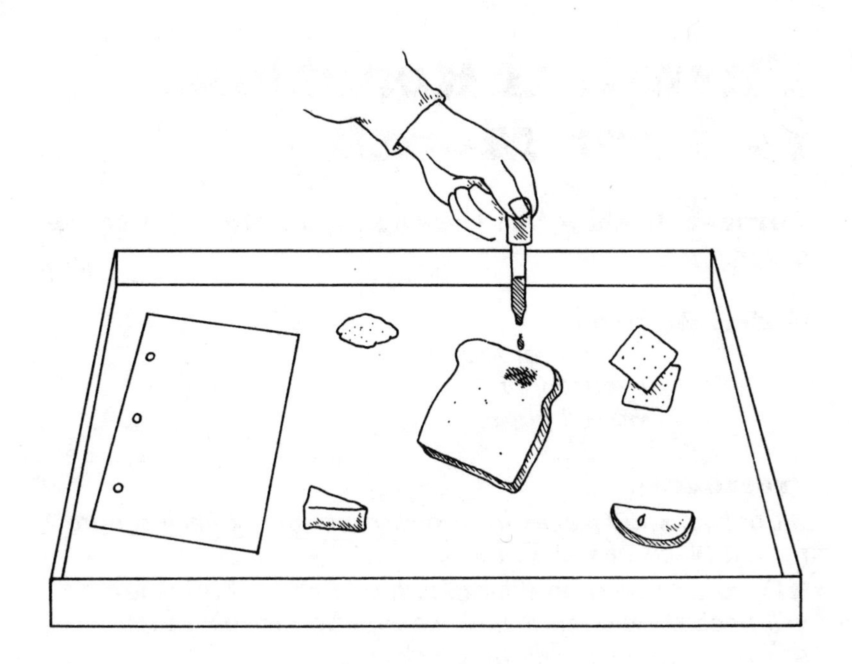

51. Chemical Reactions in Your Mouth

Purpose To show that chewing is part of a chemical reaction.

Materials *bread*
Tincture of Iodine
eyedropper
waxed paper

Procedure

- *Cut two small pieces from a slice about one-inch square from a piece of white bread.*
- *Place one piece in your mouth and chew it 30 times. It will become very mushy. Make an effort to mix as much saliva as possible with the bread.*
- *Spit the mushy bread and saliva mixture onto a piece of waxed paper.*
- *Place the second dry piece of bread on a separate piece of waxed paper.*
- *Add four drops of iodine to both bread pieces.*

Results The unchewed bread turns a dark blue-purple. The bread-saliva mixture does not turn dark.

Why? The starch in the bread combines with iodine to form an iodine-starch molecule. These molecules are blue-purple in color. Chewing the bread mixes it with saliva. The saliva chemically changes the large starch molecules to smaller sugar molecules. Sugar does not react with the iodine, thus no specific color change.

Calvin
CRACKERS
IODINE

52. Magic Writing

Purpose To write a message that magically appears.

Materials *soup bowl*
Tincture of Iodine
lemon
notebook paper
cup
art brush

Procedure

- *Pour ½ cup water into a bowl.*
- *Add 10 drops of Tincture of Iodine to the water and stir.*
- *Squeeze the juice of the lemon into the cup.*
- *Cut a section from the notebook paper. The paper must fit inside the bowl.*
- *Dip the art brush into the lemon juice and write a message on the piece of paper.*
- *Allow the juice to dry on the paper.*
- *Submerse the paper in the iodine solution in the bowl.*

Results The paper turns a blue-purple except where the message was written. The words are outlined by the dark background.

Why? The starch in the paper combines with the iodine forming iodine-starch molecules. These molecules are blue-purple in color. Vitamin C combines with iodine to form a colorless molecule. The area covered with lemon juice remains unchanged because the paper is coated with vitamin C from the lemon.

I LOV
I LOVE
WADE

53. Drinkable Iron

Purpose To test for the presence of iron in fruit juices.

Materials *1-pint glass jar*
3 tea bags
pineapple juice
apple juice
white grape juice
cranberry juice
5 clear plastic glasses
tablespoon

Procedure

■ *Make a strong tea solution by placing the tea bags in the pint jar; then fill it with hot water.*

■ *Allow the jar to stand for one hour.*

■ *Pour 4 tablespoons of each juice sample into a different glass, as shown in the illustration.*

■ *Add 4 tablespoons of tea to each glass and stir.*

■ *Allow the glasses to sit undisturbed for 20 minutes.*

■ *Carefully lift each glass and look up through the bottom of the glass. Make note of the juice that has dark particles settling on the bottom of the glass.*

■ *Allow the glasses to sit for two hours more.*

■ *Again, look for dark particles on the bottom of the glasses.*

Results Dark particles are seen in the pineapple juice after 20 minutes. Particles are seen in the cranberry and white grape after two hours. No particles form in the apple juice.

Why? A chemical change takes place which is evident by the solid particles that form. The particles are not the color of the juices which is another indication that something new has been produced. Iron in the juices combines with chemicals in the tea to form the dark particles. More particles formed in a faster time in the pineapple juice because it contains more iron. The quantity and speed of the dark particles indicates the quantity of the iron in the juice.

54. Curds and Whey

Purpose To separate milk into its solid and liquid parts.

Materials *milk*
vinegar
small baby food jar
tablespoon

Procedure

■ *Fill the jar with fresh milk.*
■ *Add 2 tablespoons of vinegar and stir.*
■ *Allow the jar to sit for two to three minutes.*

Results The milk separates into two parts; a white solid and a clear liquid.

Why? A *colloid* is a mixture of liquids and very tiny particles that are spread throughout the liquid. Milk is an example of a colloid. The solid particles in milk are evenly spread throughout the liquid. Vinegar causes the small undissolved particles to clump together, forming a solid called *curd*. The liquid portion is referred to as *whey*.

Dennis
MILK
VINEGAR

55. Limestone Deposits

Purpose To collect limestone and then, chemically remove it.

Materials *small baby food jar*
vinegar
limewater (See experiment LIMEWATER *for instructions on preparing limewater.)*

Procedure

- *Fill the jar with limewater.*
- *Leave the jar open and allow it to sit undisturbed for seven days.*
- *Pour out the limewater.*
- *Observe the white crust around the inside of the jar.*
- *Fill the jar one-half full with vinegar.*
- *Watch the changes that occur.*

Results A white crust covers the inside of the jar. The material in the white deposit reacts with the vinegar to produce bubbles. Large pieces of the crust fall away from the walls of the jar and dissolves in the vinegar. Within 5 minutes the glass touched by the vinegar is clear. The crust not touched by the vinegar remains on the glass.

Why? Carbon dioxide in the air mixes with the limewater and the white crust called limestone is formed. Limestone has the chemical name of calcium carbonate. When calcium carbonate is mixed with vinegar, a reaction takes place in which bubbles of carbon dioxide are produced.

LIMEWATE

56. A Different Form

Purpose To produce a different form of matter.

Materials *1-liter plastic soda bottle*
1 large balloon, 18 inches
1 teaspoon baking soda
3 tablespoons vinegar
cellophane tape

Procedure

■ *Pour the baking soda into the bottle.*

■ *The vinegar is to be poured into the balloon.*

■ *Attach the open end of the balloon to the mouth of the bottle. Use the tape to secure the balloon to the bottle.*

■ *Raise the balloon to allow the vinegar to pour into the bottle.*

Results The mixture starts to bubble and the balloon inflates.

Why? A chemical change occurs when the vinegar and baking soda mix together. The balloon inflates because it becomes filled with the carbon dioxide gas produced. The starting materials were in the solid and liquid form, and one of the products from the reaction is in the gas form.

Davin
VINEGAR
Amber
BAKING
SODA

5
Phase Changes

57. Colder Water

Purpose To lower the temperature of icy water.

Materials *1 small metal can*
outdoor thermometer
1 tablespoon table salt
crushed ice

Procedure

- *Fill the can with crushed ice.*
- *Cover the ice with water.*
- *Insert the thermometer.*
- *Wait 30 seconds and record the temperature of the icy water.*
- *Add 1 tablespoon of table salt to the icy water and stir very gently with the thermometer.*
- *Wait 30 seconds and record the temperature.*

Results The temperature lowers when the salt is added.

Why? It takes energy for the salt crystals to break apart into tiny particles small enough to dissolve in the water. This needed energy is obtained by removing heat from the water, which causes the water to be colder.

°C
Bobo
SALT

58. Growing Ice

Purpose To demonstrate that water expands when frozen.

Materials *1 straw*
1 small baby food jar
red or blue food coloring
permanent marking pen
clay, a piece the size of a marble

Procedure

- *Press the piece of clay against the inside bottom of the jar.*
- *Fill the jar with water.*
- *Add four or five drops of food coloring and stir.*
- *Slowly lower the straw into the colored water.*
- *Push the bottom end of the straw into the clay. The straw can now stand in a vertical position.*
- *Slowly pour all of the water out of the jar.*
- *Use the pen to mark the height of the water in the straw.*
- *Place the jar in a freezer for five hours.*

Results The height of the frozen water is above the mark.

Why? Water molecules are attracted to one another, and when they get close enough they bond or stick together. They do not stack together like flat boxes, but have spaces between them. Liquid water molecules occupy less volume because at the higher temperature the molecules are more flexible and can crowd together. As the temperature lowers, the molecules bond together to form a hexagonal structure. This ice structure is not very flexible and takes up more space than the same number of liquid water molecules.

Try This Allow the jar to stand at room temperature until the ice in the straw melts.

Results The height of the liquid water is again at the mark on the straw.

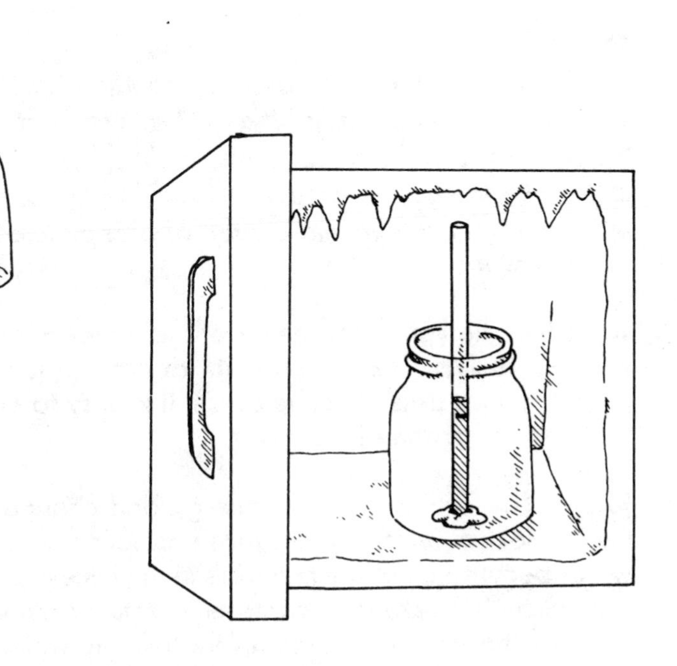

59. Frozen Orange Cubes

Purpose To determine whether orange juice will freeze like water.

Materials *orange juice*
ice tray
refrigerator

Procedure

- *Fill one-half of the ice tray with orange juice.*
- *Fill the remaining half of the ice tray with water.*
- *Set the tray in the freezer over night.*
- *Remove the frozen cubes.*
- *Carefully try to bite into a cube of orange juice and a cube of water.*

Results The liquid orange juice and water both change to solids. The frozen cube of orange juice is not as firm as is the cube of ice. It is easy to eat the cube of orange juice.

Why? The liquids both lost energy and changed from liquids to solids. Orange juice does not become as firm as the water because all of the materials in the juice are not frozen. Many liquids freeze at a lower temperature than water does. Most of the frozen material in the juice is water. The juice cube is a combination of frozen and unfrozen material which makes it easy to eat.

60. Anti-Freezing

Purpose To show that salt makes it harder for water to freeze.

Materials *2 5-oz. paper cups*
1 tablespoon table salt
marking pen

Procedure

- *Fill both cups one-half full with water.*
- *Dissolve 1 tablespoon of salt into one of the cups.*
- *Mark an S on the cup containing the salt.*
- *Set both cups in the freezer of a refrigerator.*
- *Check the cups every 30 minutes for one day, then leave the cups for 24 hours.*

Results The salty water never freezes.

Why? The salt causes the water to freeze at a lower temperature. At the freezing temperature of water, 0°C, the water molecules start linking together to form ice crystals. The salt gets in the way of this linking process, and a lower temperature is needed before the water can freeze.

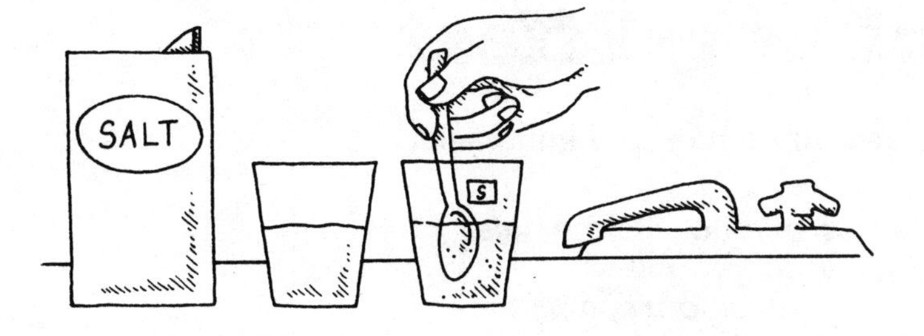
SALT
S

S

61. Chilling Effect

Purpose To cool off a thermometer.

Materials *outdoor thermometer*
cotton ball
rubbing alcohol

Procedure

■ *Lay the thermometer on a table undisturbed for three minutes; this will allow it to register the room's temperature.*

■ *Blow your breath across the thermometer bulb about 15 times.*

Results The liquid in the thermometer rises.

Procedure

■ *Moisten a cotton ball about the size of a walnut with rubbing alcohol.*

■ *Spread a thin layer of the wet cotton across the bulb of the thermometer.*

■ *Blow your breath across the wet cotton about 15 times.*

Results The liquid in the thermometer moved downward.

Why? The temperature of one's breath is about 98.6°F, which is higher than the air temperature in the room. The heat from your breath warmed the liquid in the thermometer and caused it to expand. *Expand* means that the molecules move farther apart and take up more space, thus the rise of the liquid in the thermometer.

The cooling effect of the alcohol is due to the evaporation of the alcohol around the thermometer bulb. *Evaporation* occurs when a liquid absorbs enough heat energy to change from a liquid to a gas. The alcohol takes energy away from the liquid in the thermometer bulb when it evaporates, causing the liquid to cool. Liquids contract when cooled. *Contract* means the molecules get closer together and take up less space, thus the liquid in the thermometer moves down.

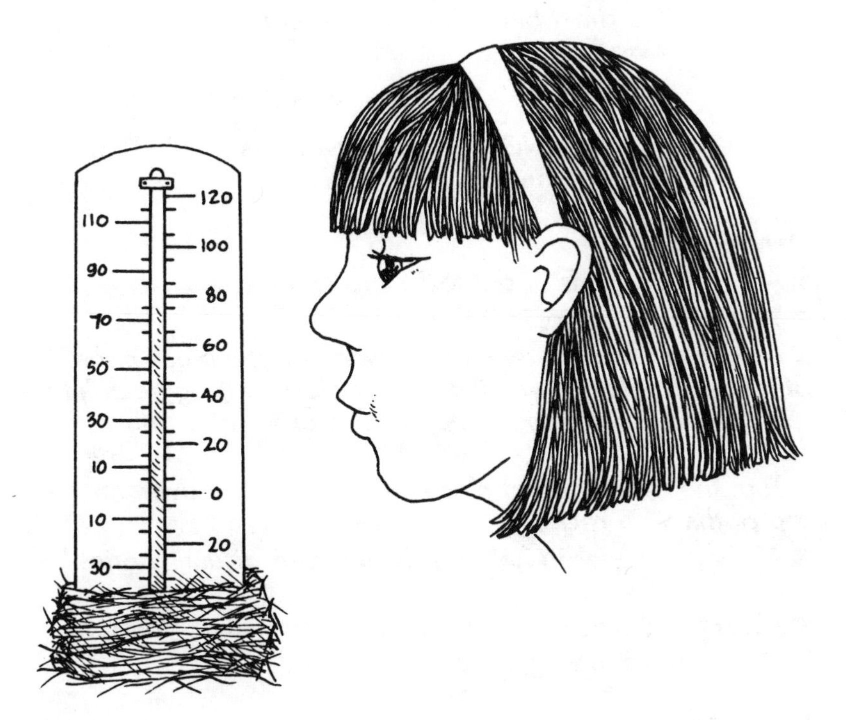

62. Crystal Ink

Purpose To produce a message written with shiny crystals.

Materials *table salt*
1 sheet black construction paper
art brush
teaspoon
stove with an oven
WARNING: Adult supervision is needed for use of the oven.

Procedure

- *Add 3 teaspoons of salt to ¼ cup water.*
- *Warm the oven to 150°F.*
- *Use an art brush to write a message on the black paper. Stir the salt solution with the brush before making each letter. It is important that this be done in order to produce a clear message.*
- *Turn the oven off and place the paper in the oven on top of the wire racks.*
- *Allow the paper to heat for five minutes or until it dries.*

Results The message appears as white, shiny crystals on a black background.

Why? The water evaporates, leaving dry salt crystals on the paper. Evaporation is the process by which a material changes from a liquid to a gas. Liquid molecules are in constant motion, moving at different speeds and in different directions. When the molecules reach the surface with enough speed, they break through and become gas molecules. Heating the paper speeds up the evaporation process.

SALT

RUSS
+
GINGER

63. Fluffy and White

Purpose To observe the growth of fluffy white crystals.

Materials *4–5 charcoal briquets*
1 tablespoon household ammonia
2 tablespoons water
1 tablespoon table salt
2 tablespoons laundry bluing
2-quart glass bowl

Procedure

- *Place the charcoal briquets in the bottom of the bowl.*
- *In a cup, mix together the ammonia, water, table salt and bluing.*
- *Pour the liquid mixture over the charcoal.*
- *Allow the bowl to sit undisturbed for 72 hours.*

Results White fluffy crystals form on top of the charcoal and some are climbing up the sides of the bowl.

Why? There are different kinds of chemicals dissolved in the water. As the water evaporates, a thin layer of crystals forms on the surface. These crystals are porous like a sponge and the liquid below moves into the openings. Water again evaporates at the surface leaving another layer of crystals. This continues, resulting in a buildup of fluffy white crystals.

Kim's
BLUING
Ginger
Ammonia
Quick Start
CHARCOAL
Bobo
SALT

64. Frosty Can

Purpose To observe the effect that salt has on the temperature of water.

Materials *crushed ice*
water
3 tablespoons of table salt
metal food can that will hold about 2 cups liquid

Procedure

- *Fill the can with crushed ice.*
- *Add 1 cup water to the can.*
- *Wait two to three minutes until water collects on the outside of the can.*
- *Add 3 tablespoons of table salt to the icy water.*
- *Gently stir the icy, salt water.*
- *Allow the can to sit for five minutes or until a thin layer of frost forms on the outside of the can.*

Results At first the outside of the can becomes covered with water. This water on the outside froze when salt was added to the icy water.

Why? Air contains molecules of water in gas form. This gas cools when it touches the chilled can and changes to liquid water. The salt lowers the temperature of the icy water, which causes the temperature of the can to be lowered. The drops of water on the outside freeze and form a frosty coating around the can.

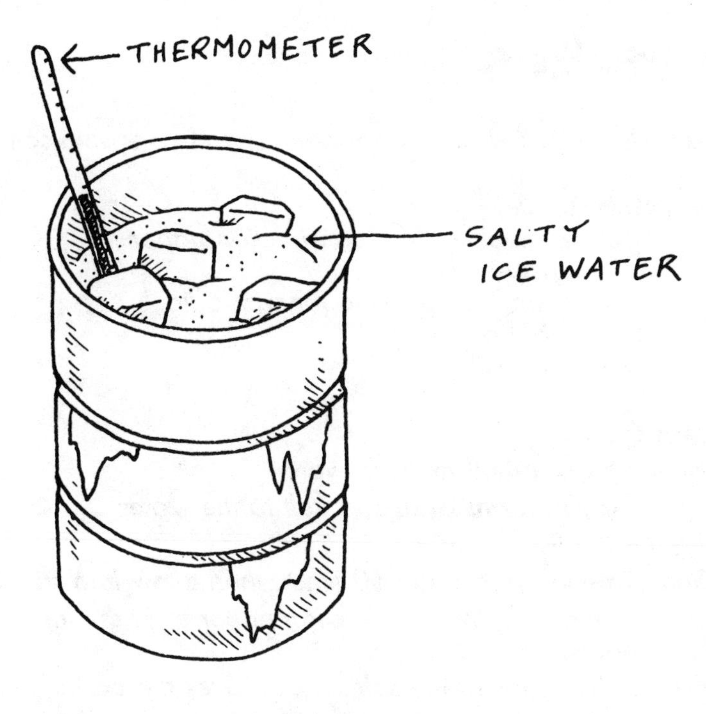
THERMOMETER
SALTY
ICE WATER

65. Needles

Purpose To grow needle-shaped crystals of epsom salt.

Materials *saucer*
1 sheet of dark construction paper
epsom salt
1 small baby food jar with a lid
tablespoon
scissors

Procedure

- *Fill the jar one-half full with water.*
- *Add 2 tablespoons of epsom salt to the water.*
- *Secure the lid.*
- *Shake the jar vigorously 60 times, then allow it to stand.*
- *Cut a circle from the construction paper to fit the inside of the saucer.*
- *Pour a thin layer of the salt solution over the paper. Try not to pour out the undissolved salt.*
- *Place the saucer in a warm place and wait several days.*

Results Long, slender, needle-shaped crystals form on the paper.

Why? Epsom salt crystals are long and slender. The particles in the box have been crushed for packaging and do not have a slender shape. As the water evaporates from the solution, small, unseen crystals start to stack together. Further evaporation increases the building process and long, needle-shaped crystals are produced.

Jenny
EPSOM
SALT

66. Lace

Purpose To grow a layer of lacy salt crystals.

Materials *3 tablespoons table salt*
cup
tall, slender, clear jar
black construction paper
scissors

Procedure

■ *Pour ½ cup water into the jar.*

■ *Add the salt and stir.*

■ *Cut a ½-inch strip from the construction paper. The height of the paper should be about one-half the height of the jar.*

■ *Stand the paper strip against the inside of the jar.*

■ *Place the jar in a visible place where it will be undisturbed.*

■ *Allow the jar to sit for three to four weeks. Observe it daily.*

Results Lacy crystals may be seen at the top of the paper after several days. More lace develops the longer the jar sits.

Why? The salty water moves up the paper and onto the glass where it spreads out. The water evaporates leaving microscopic bits of salt on the glass. This continues until visible crystals of salt are seen. The water continues to evaporate at the edge, producing layers of lacy crystals around the inside of the jar.

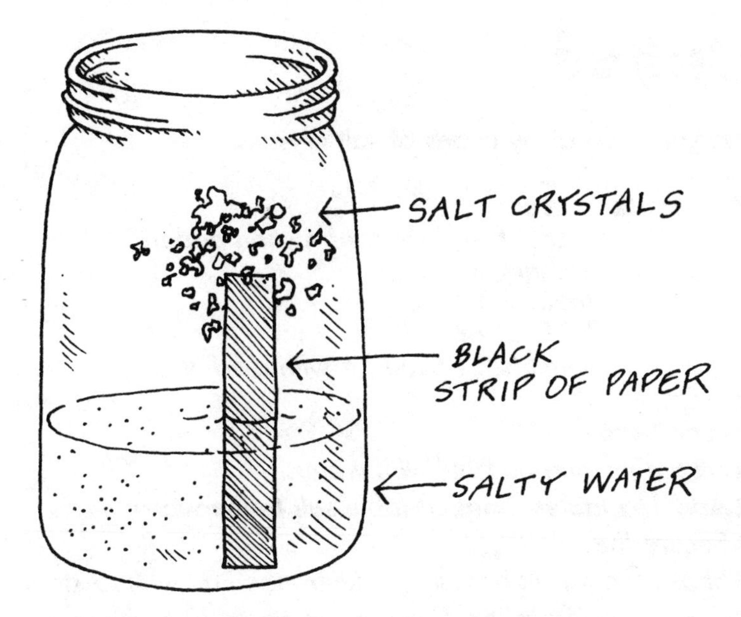
SALT CRYSTALS
BLACK
STRIP OF PAPER
SALTY WATER

67. Cubes

Purpose To grow cubes of salt crystals.

Materials *saucer*
1 sheet of dark construction paper
scissors
table salt
tablespoon
small baby food jar with a lid

Procedure

■ *Fill the jar one-half full with water.*

■ *Add 1 ½ tablespoons of table salt to the water.*

■ *Secure the lid.*

■ *Shake the jar vigorously 30 times, then allow it to stand.*

■ *Cut a circle from the construction paper to fit the inside of the saucer.*

■ *Pour a thin layer of the salt solution over the paper. Try not to pour any undissolved salt onto the paper.*

■ *Place the sauce in a warm place and wait several days.*

■ *Observe the paper daily.*

Results Small, white, cubic crystals form on the paper and increase in size every day.

Why? As the water evaporates, dry salt is deposited on the paper. Table salt crystals have a cubic shape. The tiny unseen crystals are deposited first and, as more water evaporates, the crystals stack until they are large enough to be seen.

68. Plaster Block

Purpose To observe a phase change due to the addition of water.

Materials *⅓ cup Plaster of Paris*
tablespoon
paper cup
plastic spoon

Procedure

■ *Pour ⅓ cup Plaster of Paris into the paper cup.*

■ *Add 3 tablespoons of water and stir with the plastic spoon.*

Note: *Be careful not to put any of the plaster in the sink as it can clog the drain. Throw the plastic spoon away.*

■ *Squeeze the cup gently and observe the results every 20 minutes.*

Results A very thick liquid is produced at first. The expected results after:

a. 20 min.: Water collects on top.
b. 40 min.: The liquid is thicker.
c. 60 min.: The liquid is very thick and sticks to the sides of the cup.
d. 80 min.: The liquid is becoming firmer.
e. 120 min.: It is no longer a liquid. The material has solidified, but still feels moist.
f. 140 min.: The block is firm.

During these changes the cup feels warm.

Why? Plaster of Paris is made by grinding a clear, shiny crystal called gypsum into a powder. The powder is heated to remove all of its moisture. This dry powder changes back into a solid when water is added, but it never looks clear and shiny again. Heat is given off during this phase change.

PLASTER
of
PARIS

6

Solutions

69. Streamers of Color

Purpose To observe the dissolving of a solute in a solvent.

Materials *clear drinking glass*
powdered fruit drink
flat toothpick

Procedure

■ *Fill the glass with water.*

■ *Select a flavor of powdered fruit drink that has a dark color such as cherry, grape, raspberry, etc.*

■ *Use the wide end of a flat toothpick to pick up a scoop of the powdered drink.*

■ *Gently shake the powder over the glass of water.*

■ *Observe from the side of the glass.*

■ *Continue to add the powder until the water becomes completely colored.*

Results Streamers of color precipitate down through the water. *Precipitate* means to fall downward.

Why? The crystals dissolve in the water as they fall. *Dissolving* means that a substance breaks apart into smaller and smaller particles and spreads out evenly throughout the solvent. The dissolving material, the *solute,* is the powdered crystals and the *solvent* is the water. The combination of a solute and a solvent produces a liquid solution.

70. Tasty Solution

Purpose To determine the fastest way to dissolve candy.

Materials *3 bite-sized pieces of soft candy.*

Procedure

■ *Place one of the candy pieces in your mouth.* DO NOT *chew, and* DO NOT *move your tongue around.*

■ *Record the time it takes for this candy piece to dissolve.*

■ *Place a second candy piece in your mouth.* DO *move the candy back and forth with your tongue, but* DO NOT *chew.*

■ *Record the time it takes to dissolve this candy piece.*

■ *Place the third piece of candy in your mouth. DO move the candy back and forth with your tongue as you chew.*

■ *Record the time to dissolve this third piece of candy.*

Results Moving the candy around and chewing it decreased the time necessary for dissolving.

Why? The candy dissolves in the saliva in your mouth to form a liquid solution. Solutions contain two parts, a solute and a solvent. The solvent is the saliva and the solute is the candy. The solute dissolves by spreading out evenly throughout the solvent. The candy can quickly dissolve when it is crushed by chewing and stirred by moving it around with the tongue.

Krysti
CANDY
TIME

71. Speedy Soup

Purpose To make a quick, tasty cup of soup.

Materials *2 bouillon cubes*
2 cups
hot and cold water

Procedure

- *Fill one cup with cold water from the faucet.*
- *Add one bouillon cube.*
- *Allow this cup to sit undisturbed while the second cup is prepared.*
- *Fill the second cup with hot water from the faucet.*
- *Add one bouillon cube to the water and stir.*

Results The solid cube dissolved more quickly when placed in hot water and stirred.

Why? Dissolving means that the solute moves evenly throughout the solvent. The bouillon cube is the solute and the water the solvent. Heat causes the molecules of water to move faster, thus the water molecules hit against the cube causing pieces to break off. Stirring increases the breaking process. The cube will finally dissolve in the cold water, but it takes a much longer period of time. Stirring the cold water will help speed up the dissolving.

72. Rainbow Effect

Purpose To observe the separation of colors in ink.

Materials *green and black water-soluble pens*
coffee filter
saucer
paper clip

Procedure

- *Fold the coffee filter in half.*
- *Fold it in half again.*
- *Make a dark green mark about one inch from the rounded edge of the folded filter.*
- *Make a second mark with the black marker about one inch from the rounded edge. The two marks are not to touch each other, but need to be on the same side.*
- *Secure the edge of the filter with the paper clip so that a cone is formed.*
- *Fill the saucer with water.*
- *Place the rounded edge of the cone in the water.*
- *Allow the paper to stand undisturbed for one hour.*

Results It takes about one hour for the colors to separate. A trail of blue, yellow, and purple is seen from the black mark, and the green mark produces a trail of blue and yellow.

Why? Black and green are combinations of other colors. As the water rises in the paper the ink dissolves in it. Some of the colors rise to different heights depending on the weight of the chemicals producing the color. The lighter weight chemicals move with the water to the top of the paper.

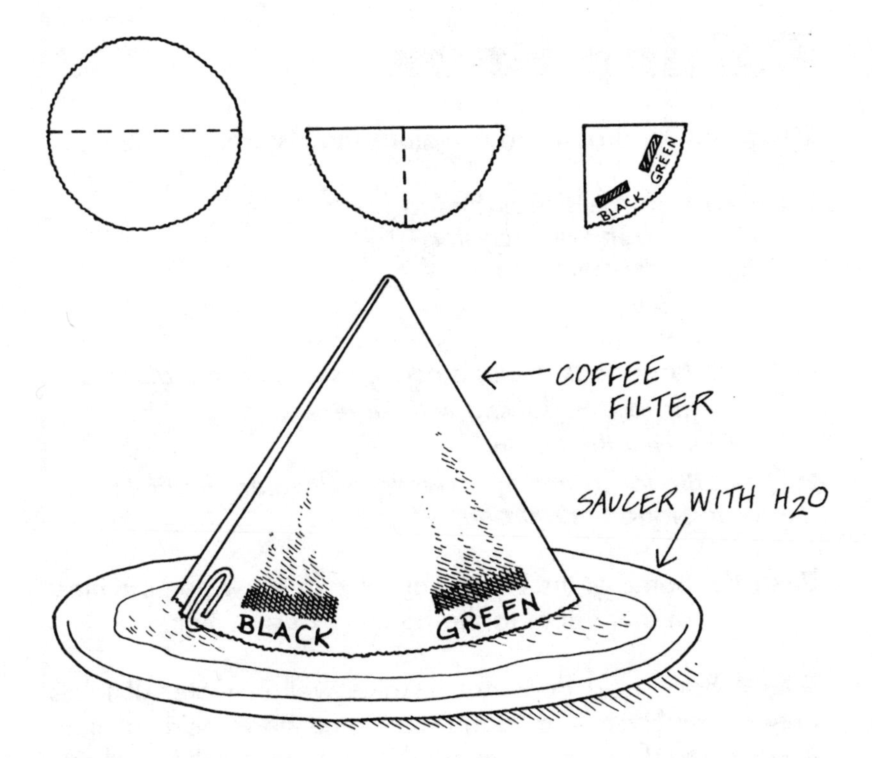
GREEN
BLACK
COFFEE FILTER
SAUCER WITH H_2O
BLACK
GREEN

73. Falling Snow

Purpose To produce a miniature snow storm.

Materials *large-size baby food jar with lid*
boric acid crystals
teaspoon

Procedure

■ *Pour 5 teaspoons of boric acid crystals into the glass jar.*
■ *Fill the jar to overflowing with water.*
■ *Tightly screw the lid on.*
■ *Shake the jar to mix the crystals and water, then allow the jar to stand undisturbed.*

Results Some of the crystals dissolve in the water, but most of them float to the bottom like snowflakes.

Why? Boric acid does not dissolve well in water. It takes only a few crystals to form a saturated boric acid solution. A *saturated* solution is one in which no more solute will dissolve. Shaking the jar causes the undissolved crystals to float around and then gravity pulls them to the bottom of the jar.

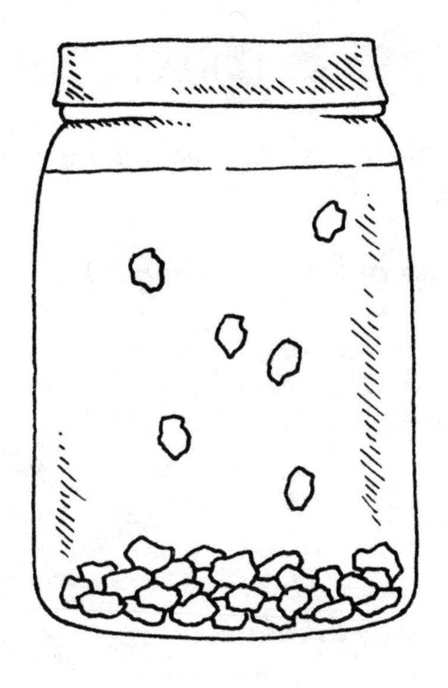

74. Floating Spheres

Purpose To float spheres of colored water between layers of water and oil.

Materials *¼ cup of liquid cooking oil*
¼ cup water
1-pint glass jar
blue food coloring
eyedropper

Procedure

- *Pour the water in the jar.*
- *Slowly add the liquid oil.*
- *Use the eyedropper to add five drops of food coloring to the jar.*
- *While holding the jar at eye level, look at the underside of the oil's surface.*
- *Use a pencil to push the drops of coloring into the water.*

Results Two separate layers form. The oil floats on top of the water. The balls of food coloring float just beneath the surface of the oil. Some of the colored balls sink and sit just above the surface of the water. As the colored balls touch the water they immediately break apart and dissolve in the water.

Why? Oil and water are immiscible. *Immiscible* means they do not mix and will separate into layers.

Food coloring does not dissolve in oil and will float if the drops are small enough. The oil surrounding the balls of color prevent them from touching the water. Pushing the balls through the oil allows them to touch and dissolve in the water.

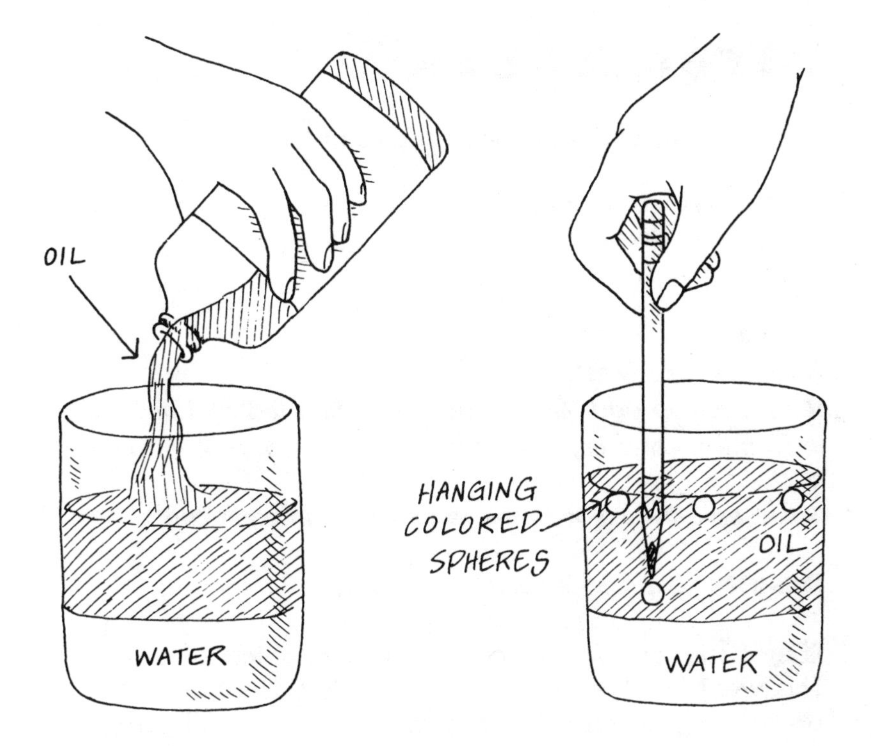
OIL
WATER
HANGING
COLORED
SPHERES
OIL
WATER

75. Strengths?

Purpose To compare the strengths of tea.

Materials *instant tea*
teaspoon
2 cups

Procedure

■ *Fill the cups with water.*

■ *To one cup add ¼ teaspoon of instant tea and stir.*

■ *To the second cup add 1 heaping teaspoon of instant tea and stir.*

■ *Observe the color of the tea solution in each cup.*

Results One of the solutions is lighter in color.

Why? The solution with the least amount of tea is less strong. This weak solution is said to be *diluted.* The darker tea is stronger. A strong solution is more concentrated. *Concentrated* means that more solute is being dissolved. All solutions are made of a solute and a solvent. The solute is the substance being dissolved in the liquid. In these tea solutions, tea is the solute that is dissolved in the water solvent.

INSTANT TEA
INSTANT TEA
DILUTE, WEAK, LIGHT
CONCENTRATED, STRONG, DARK

76. **Spinning**

Purpose To separate the floating parts of a suspension by spinning.

Materials *hammer*
nail
clear drinking glass
1-lb. metal coffee can
cotton string
2 tablespoons flour

Procedure

■ *Use the hammer and nail to make two holes across from each other beneath the top rim of the can.*

■ *Tie the ends of a two-foot piece of string in these holes.*

■ *Fill the can one-half full with water.*

■ *Stir the flour into the water.*

■ *Carry the can and the empty glass outside.*

■ *Hold the string and swing the can around 15 times.*

■ *Pour a small amount of the liquid into the empty glass. If it looks cloudy, swing 10 more times.*

■ *Continue to swing and test for cloudiness until the liquid stops changing.*

Results The solution clears.

Why? Flour and water is a mixture of a liquid and solid. The mixed parts do not dissolve and the solid particles settle to the bottom. Spinning speeds this settling process. Spinning produces a strong outward force. This force pushes the floating flour particles to the bottom of the can.

77. Layering

Purpose To observe layering of undissolved materials.

Materials *2 tablespoons flour*
2 tablespoons of any large dried bean
quart glass jar with a lid

Procedure

- *Place the beans and flour in the jar.*
- *Fill the jar with water.*
- *Screw the lid on tightly.*
- *Shake the jar to mix all the materials thoroughly.*
- *Let the jar stand undisturbed for 20 minutes.*
- *Observe.*

Results The beans settle first with a fine layer of flour on top.

Why? The beans and flour are not soluble in the water. As soon as the shaking stops, gravity starts pulling the materials down. The heavier beans settle first. The tiny flour particles remain suspended in the water for a few minutes, but finally are pulled to the bottom of the jar. The mixing of materials that do not combine is called a *suspension*. Water in a fast-flowing stream forms a suspension by picking up rocks and soil that temporarily stay suspended in the moving water, but the materials settle out in layers on the stream bed when the water's speed is reduced.

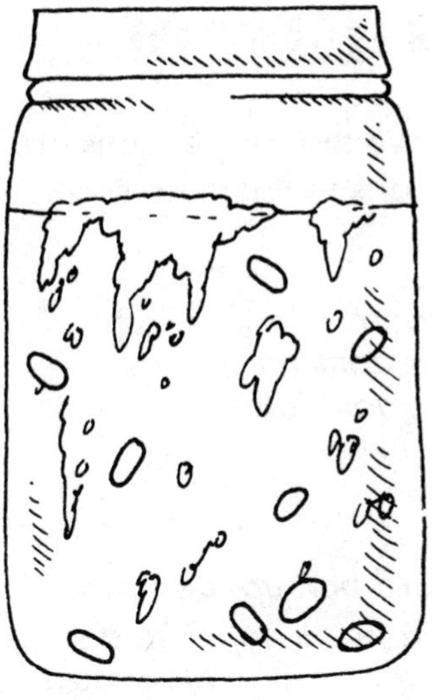

78. Tyndall Effect

Purpose To observe that suspensions are cloudy and contain solid floating parts that can be seen.

Materials *scissors*
cardboard box
2 clear drinking glasses
1 teaspoon flour
flashlight

Procedure

- *Turn the cardboard box upside down.*
- *Use the point of a pencil to make a small hole in the end of the box. The height of the hole should be one-half the height of the glass being used.*
- *Cut a one-inch square viewing hole in the front of the box. The hole must be about three inches from the corner of the box and as high as the small round hole on the side.*
- *Fill the glasses three-quarters full with water.*
- *Add 1 teaspoon of flour to one of the glasses with water and stir.*
- *Place the glass containing water and flour under the box. Position the glass so that it is in front of the viewing hole.*
- *Hold the flash light near the small hole.*
- *Observe the effect that the liquid has on the light rays.*
- *Put the glass containing only water under the box.*
- *Shine the light through the hole and observe the effect that water has on the light rays.*

Results The mixture of flour and water looked cloudy. Tiny particles of flour could be seen floating in the water. The glass of water had no effect on the light rays, they passed through the water unchanged.

Why? Flour and water form a suspension. A suspension has tiny particles floating in the liquid. The particles stay suspended until gravity pulls them down. The suspended particles stop some of the light rays. Light hits the bits of flour floating in the water and is reflected. *Reflect* means to bounce back. There is nothing in the water to reflect the light. Reflection of light by suspended particles is called the *Tyndall Effect,* named after the British scientist, John Tyndall.

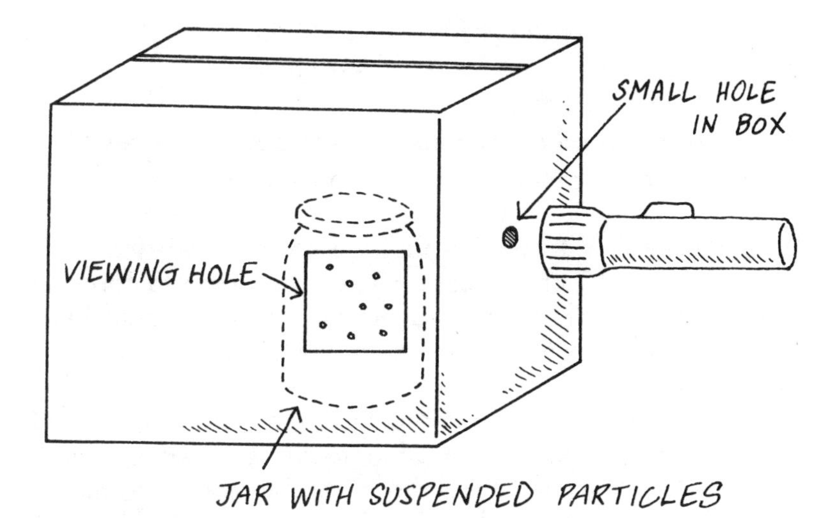

79. Immiscible

Purpose To observe the separation of an emulsion.

Materials *¼ cup liquid oil*
½ cup water
blue food coloring
1-quart glass jar with a lid

Procedure

- *Pour the water into the jar.*
- *Add five drops of the food coloring and stir.*
- *Slowly add the liquid oil.*
- *Secure the lid and shake the jar vigorously ten times.*
- *Put the jar on a table and observe what happens.*

Results At first it appears that the liquids have dissolved, but in only seconds three layers start to form. In only minutes two layers are present. Liquid bubbles are present in all the layers.

Why? Oil and water are *immiscible,* meaning they do not mix. A combination of immiscible liquids is called an *emulsion.* Shaking the jar causes the oil and water to be mixed together, but they immediately start to separate. The heavier water sinks to the bottom carrying with it trapped bubbles of oil. The center layer has an even distribution of oil and water, making it heavier than the oil but lighter than water. The top layer is mostly oil with trapped bubbles of water in it. It takes about eight hours for all of the oil to rise and all of the water to sink. Since only the water is colored, the food coloring has to be water soluble.

OIL

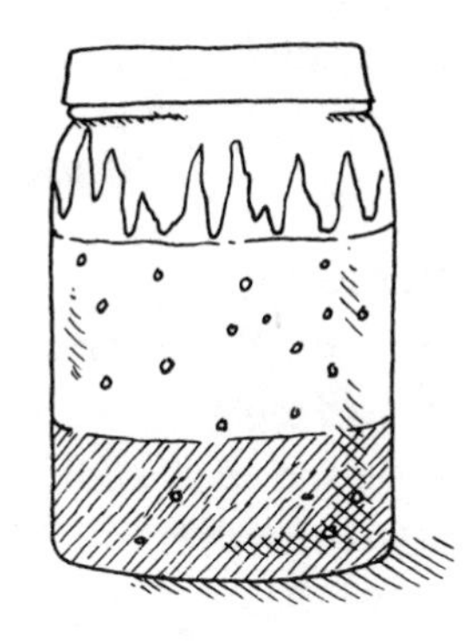

80. **Dilution**

Purpose To observe a color change as a solution becomes more dilute.

Materials *1-gallon glass jar*
1 cup
red food coloring

Procedure

- *Pour ½ cup of water into the gallon jar.*
- *Add and stir in one drop of food coloring.*
- *Add one cup of water at a time to the jar until the red color disappears.*

Results It takes about seven cups of clear water to make the red color disappear.

Why? The red is visible at first because the molecules of red color are close enough together to be seen. As clean water is added, the color molecules continues to spread evenly throughout the water. They finally get far enough apart to become invisible because of their small size.

81. Spicy Perfume

Purpose To make a bottle of spicy perfume.

Materials *small baby food jar with lid*
rubbing alcohol
15 whole cloves

Procedure

- *Place the whole cloves in the jar.*
- *Fill the jar one-half full with the rubbing alcohol.*
- *Secure the lid and allow the jar to set for seven days.*
- *Use your finger to dab a few drops of the alcohol on your wrist.*
- *Allow the alcohol to evaporate, then smell your wrist.*

Results The skin has a faint, spicy smell.

Why? The alcohol dissolves the aromatic oil in the cloves. When the alcohol evaporates from the wrist, the scented oil is left on the skin. Perfumes are made by dissolving oils from flowers and other aromatic materials in alcohol.

ROBERT'S RUBBING ALCOHOL
CAROL'S CLOVES

7 Heat

82. Smoke Rings

Purpose To observe the downward flow of cold colored water through warmer clear water.

Materials *1 large-mouthed, clear, glass, quart jar*
red food coloring
small baby food jar
6-inch square of aluminum foil
rubber band
pencil
1 ice cube

Procedure

■ *Place the ice cube in the baby food jar. Fill the jar with cold water.*

■ *Fill the quart jar to within an inch of the top with hot water from the faucet.*

■ *Remove the ice cube from the baby food jar. Add and stir in six to seven drops of food coloring.*

■ *Cover the mouth of the baby food jar with aluminum foil. Use the rubber band to secure the foil around the mouth of the jar.*

■ *Use the point of the pencil to make a small hole in the aluminum foil.*

■ *Quickly turn the baby food jar upside down and hold it so that the hole is just beneath the surface of the hot water.*

■ *Slowly and gently tap the bottom of the baby food jar with the eraser of the pencil or your finger.*

Results The cold colored water flows downward. The tapping causes the colored water to come out in spurts, producing smoke-like rings of color in the warm clear water.

Why? Cold water weighs more than warm water because the cold water molecules are closer together. The molecules of water, like all matter, are spaced closer together when cold and farther apart when heated. The food coloring has little or no effect on the weight. Since the cold water is heavier, it sinks down through the lighter warmer water.

83. Puff Signals

Purpose To observe the movement of hot, colored water through cooler clear water.

Materials *2 large-mouthed, clear, glass, quart-jars*
red food coloring
small baby food jar
6-inch square of aluminum foil
rubber band
pencil
4 or 5 ice cubes

Procedure

■ *Place the ice cubes in one of the quart jars. Fill the jar with cold water.*

■ *Fill the baby food jar to over-flowing with hot tap water. Add and stir in six or seven drops of food coloring.*

■ *Cover the mouth of the baby jar with aluminum foil. Use the rubber band to secure the foil around the mouth of the jar.*

■ *Stand the baby food jar inside the empty wide-mouthed jar.*

■ *Remove any unmelted ice cubes and pour the chilled water into the container with the baby food jar. Completely fill the jar with the cold water.*

■ *Use the point of the pencil to make a small hole in the aluminum foil.*

■ *Slowly and gently tap the foil with the eraser of the pencil.*

Results The hot, colored water puffs upward like smoke rings.

Why? Water molecules, like all matter, are spaced closer together when cold and farther apart when heated. The colored hot water weighs less than the colder clear water because of this spacing. The lighter hot water rises to the top of the heavier chilled water.

Try This Punch a second hole into the foil.

Results A stream of colored water starts and continues to flow out of the jar.

Why? The cold water sinks into one of the holes, pushing the lighter hot water out.

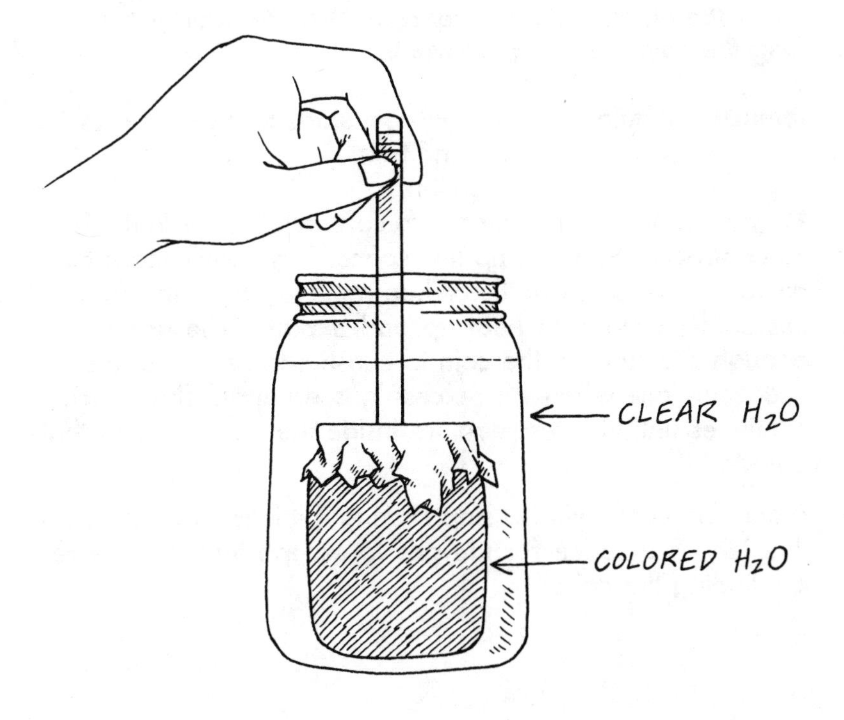

84. Clicking Coin

Purpose To observe the effects of expanding gas.

Materials *2-liter soda bottle*
quarter
cup of water

Procedure

■ *Place the empty, uncapped soda bottle in the freezer for five minutes.*

■ *Remove the bottle from the freezer and* immediately *cover the mouth with the wet coin. Wet the quarter by dipping the coin in the cup of water.*

Results Within seconds, the coin starts to make a clicking sound as it rises and falls.

Why? Cooling causes matter to contract. The air in the bottle contracts and takes up less space. This allows more cold air to flow into the bottle. When removed from the freezer this cold air starts to heat up and expand. The gas exerts enough pressure on the coin to cause it to rise on one side. The coin falls when the excess gas escapes. This process continues until the temperature inside the bottle equals that outside.

Note: The coin will also stop clicking if it falls into a position that leaves a space for the gas to escape through. Try repositioning the coin.

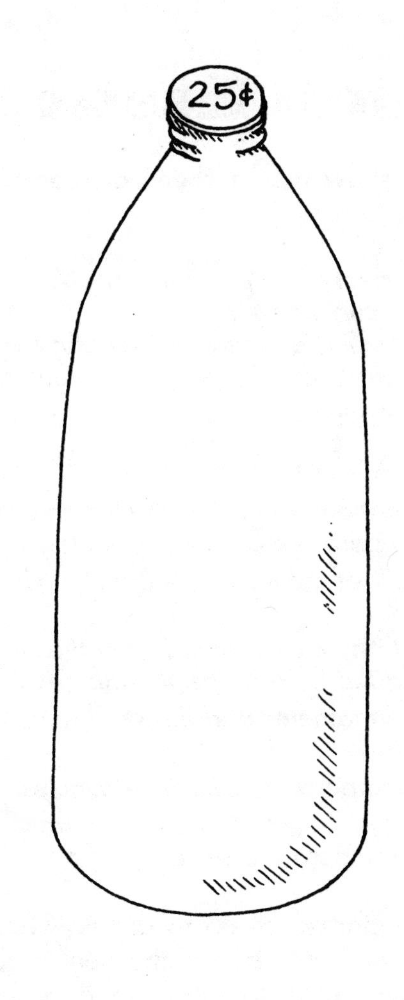
25¢

85. Chemical Heating

Purpose To show that a chemical reaction can produce heat.

Materials *1 steel wool pad without soap*
¼ cup vinegar
cooking or outdoor thermometer
1 jar with lid (The thermometer must fit inside the closed jar.)

Procedure

- *Place the thermometer inside the jar and close the lid. Record the temperature after five minutes.*
- *Soak one-half of the steel wool pad in vinegar for one or two minutes.*
- *Squeeze out any excess liquid from the steel wool and wrap it around the bulb of the thermometer.*
- *Place the thermometer and the steel wool inside the jar. Close the lid.*
- *Record the temperature after five minutes.*

Results The temperature rises.

Why? The vinegar removes any protective coating from the steel wool, allowing the iron in the steel to rust. Rusting is a slow combination of iron with oxygen, and heat energy is always released. The heat released by the rusting of the iron causes the liquid in the thermometer to expand and rise in the thermometer tube.

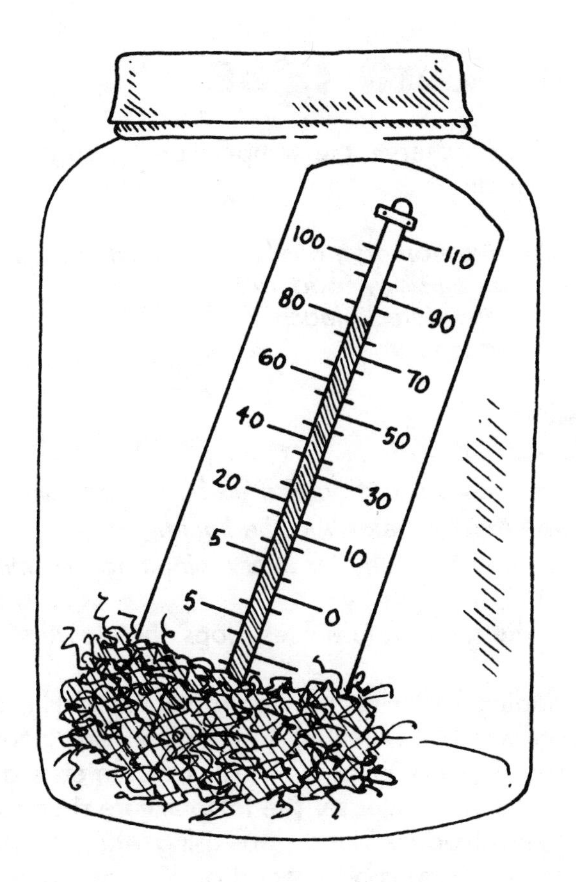

100
110
80
90
60
70
40
50
20
30
5
10
5
0

86. Heat Changes

Purpose To observe the temperature changes during a chemical reaction.

Materials *thermometer, outdoor or cooking type*
small baby food jar
powdered bleach
teaspoon

Procedure

- *Fill the jar with water.*
- *Add one teaspoon of powdered bleach and stir.*
- *Insert the thermometer into the liquid.*
- *Observe the thermometer every minute for 10 minutes.*

Results The temperature rises, stops, then lowers.

Why? Adding water to powdered bleach starts a chemical change in which oxygen is slowly given off. Heat is also released during this change. The thermometer is a way of observing this heat release. The temperature rises as long as heat is being produced and stops rising when the heat production stops. There is a lowering of the temperature as the heat in the liquid is lost to the air in the room. The temperature of the liquid will finally become the same as that of the air in the room.

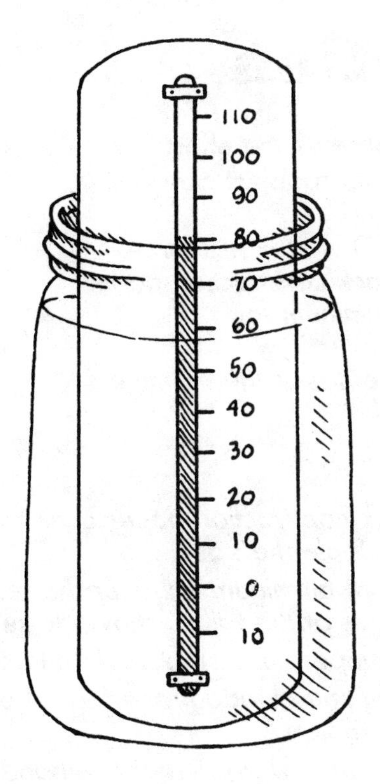
110
100
90
80
70
60
50
40
30
20
10
0
10

87. Radiation

Purpose To observe the effect that color has on the amount of radiation that an object absorbs.

Materials *100-watt light source*
black construction paper
aluminum foil
stapler
2 outdoor thermometers
ruler

Procedure

■ *Fold the black construction paper over the thermometer as shown and staple the sides.*

■ *Fold a piece of aluminum foil over the second thermometer. Fold the sides of the foil as shown to secure them.*

■ *Record the temperature on both thermometers.*

■ *Place the light source about one foot above the pouches with the thermometers.*

■ *Turn the light on and observe the temperature readings for 10 minutes.*

Results The temperature reading is higher on the thermometer in the black pouch.

Why? Black objects absorb all of the light waves. Since none of the waves of light are reflected back to the viewer the object looks black. This absorption of the waves of energy causes the object's temperature to rise. The aluminum foil does not absorb very many of the light waves, thus, its temperature is lower. Spring and summer clothes are usually light in color, so the wearer stays cooler.

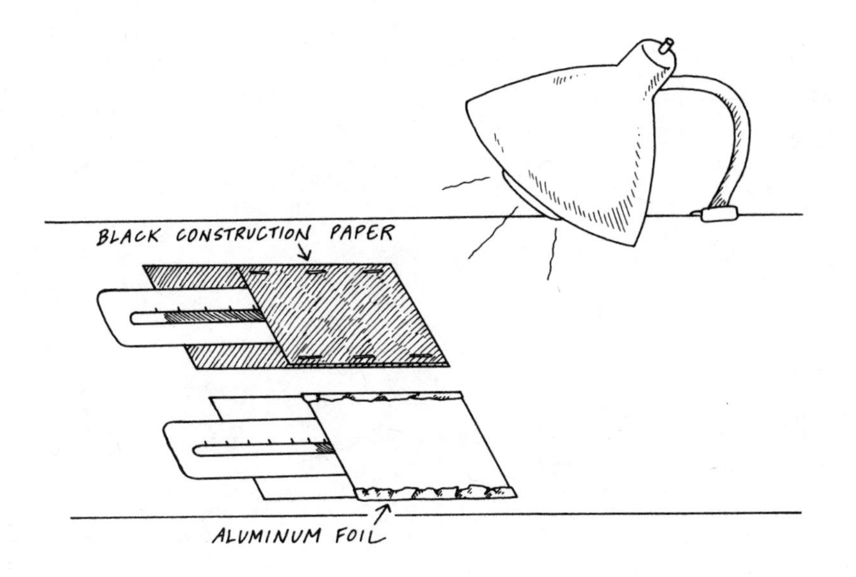
BLACK CONSTRUCTION PAPER
ALUMINUM FOIL

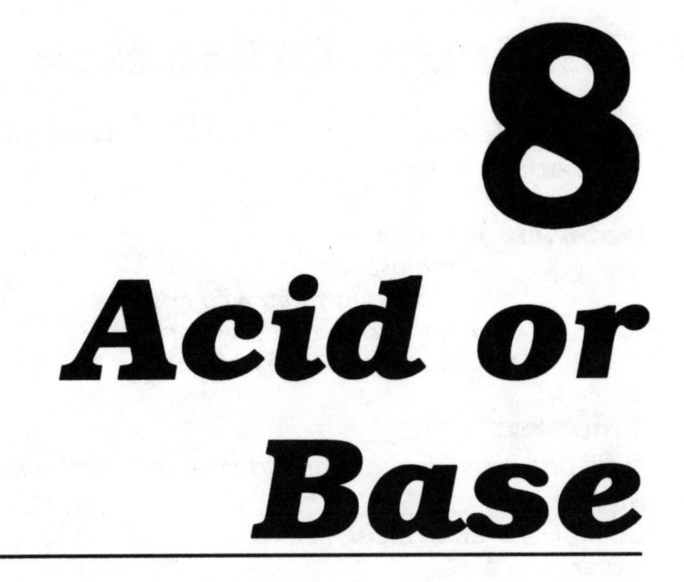

8 Acid or Base

88. Cabbage Indicator

Purpose To make a solution that will indicate the presence of an acid or a base.

Materials *tea strainer*
1 tablespoon
2 glass quart jars with lids
1 quart of distilled water
uncooked purple cabbage

Procedure

■ *Fill one jar with cabbage leaves that have been torn into small pieces.*

Caution: *Parental assistance may be needed to heat the water.*

■ *Heat the distilled water to boiling, and fill the jar containing the pieces of cabbage with the hot water.*

■ *Allow the jar to stand until the water cools to room temperature.*

■ *Pour the cooled cabbage solution through a tea strainer into the second quart jar. Discard the cabbage leaves.*

■ *Store the cabbage juice in a refrigerator until needed.*

Results After standing, the water covering the cabbage leaves turns blue.

Why? The hot water dissolves the colored chemicals in the cabbage. These colored chemicals turn red when mixed with an acid, and a base will produce a green color. Cabbage juice can be used to test for the presence of two different kinds of chemicals, *acid* and *base*.

89. Cabbage Paper

Purpose To make a paper indicator that can be used to test for an acid or a base.

Materials *coffee filters*
cabbage indicator juice (Prepared in experiment CABBAGE INDICATOR)
cookie sheet
quart bowl
scissors
zip-lock plastic bag

Procedure

■ *Pour 1 cup of cabbage juice into the bowl.*

■ *Dip one piece of filter paper into the cabbage juice.*

■ *Place the wet paper on the cookie sheet.*

■ *Continue wetting the filter paper until the cookie sheet is covered with the papers.*

■ *Allow the papers to dry.*

■ *Cut half of the dry papers into strips about one-half inch by three inches. Store the dry strips and the large papers in a zip-lock plastic bag.*

■ *The papers will be used to test for an acid or base. See experiment* ACID-BASE TESTING *for instructions on using the testing papers.*

Results A pale blue testing paper is produced.

Why? Juice extracted from purple cabbage has a bluish color. Allowing the water to evaporate from the juice leaves a pale blue chemical on the paper that changes colors when touched to an acid or a base.

CABBAGE
JUICE

90. Acid-Base Testing

Purpose To use cabbage paper to test for the presence of an acid or base.

Materials *1 cabbage paper strip (prepared in experiment* CABBAGE PAPER)
cookie sheet
1 sheet of notebook paper
2 eyedroppers
vinegar
household ammonia
2 small baby food jars

Procedure

■ *Fill one of the small jars one-quarter full with vinegar and place an eyedropper in it.*

■ *Fill the second jar one-quarter full with ammonia and place an eyedropper in the jar.*

■ *Place the notebook paper on the cookie sheet.*

■ *Lay the piece of cabbage testing paper on top of the notebook paper.*

■ *On one end of the cabbage paper place two drops of vinegar.*

■ *Add 2 drops of ammonia to the opposite end of the cabbage paper.*

Results Ammonia turns the paper green and vinegar produces a pink color.

Why? Cabbage testing paper is used to test for the presence of acids or bases. The chemicals in the cabbage juice always produce the same color changes. Bases change the paper to green, and acids produce a pink-to-red color. The cabbage paper in this lab indicates that ammonia is a basic chemical and that vinegar is acidic.

VINEGAR
AMMONIA

91. A or B

Purpose To test many different substances at one time for the presence of an acid or base.

Materials *2 eyedroppers*
1 large sheet of cabbage testing paper (prepared in experiment CABBAGE PAPER)
1 sheet notebook paper
cookie sheet
pencil
lemon
grapefruit
orange
household ammonia

Procedure

■ *Place the notebook paper on the cookie sheet.*

■ *Lay the cabbage paper on top of the notebook paper.*

■ *Use a pencil to write the names of the testing materials on the notebook paper.*

■ *Squeeze two drops of lemon juice on the cabbage paper next to the word "lemon."*

■ *Squeeze drops of grapefruit juice and orange juice on the testing paper next to their names.*

■ *Use an eyedropper to place two drops of ammonia on the testing paper.*

■ *Use a clean eyedropper to place the two drops of pickle juice on the paper.*

Results Ammonia turns the cabbage paper green. All of the remaining liquids produce a pink-to-red color.

Why? Bases turn cabbage testing paper green, and acids produce a pink-to-red color. Ammonia is basic and the other liquids are acidic. Citric acid is present in the fruit. Pickle juice contains vinegar, which has the chemical name of acetic acid.

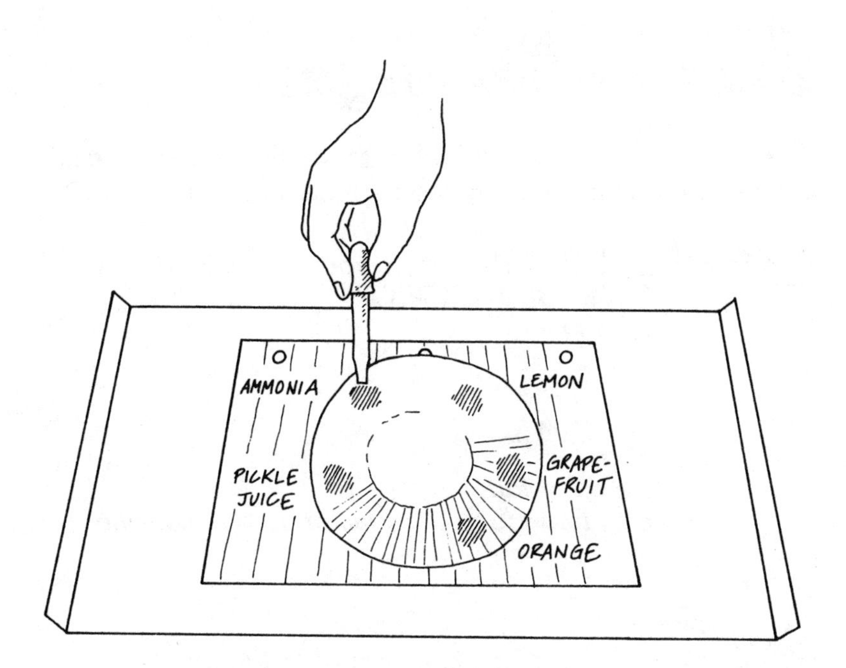
AMMONIA
LEMON
PICKLE
JUICE
GRAPE-
FRUIT
ORANGE

92. Strong-Stronger

Purpose To observe the color effect that different acid concentrations have on the cabbage testing solution.

Materials *cabbage indicator (prepared in experiment* CABBAGE INDICATOR)
scissors
filter paper
cookie sheet
teaspoon
alum
cream of tartar
Fruit Fresh (a fruit protector used in canning and freezing)

Procedure

- *Place ½ teaspoon of alum, cream of tartar, and Fruit Fresh on the cookie sheet. Space the powders about three inches apart.*
- *Cut three strips, about one-half inch by three inches, from the filter paper.*
- *Dip the end of one of the filter strips in the cabbage solution. Lay the wet end over the mound of alum.*
- *Wet a second filter strip with cabbage juice and lay over the cream of tartar.*
- *The third filter strip is to be wet with the cabbage juice and placed over the Fruit Fresh.*
- *Wait five minutes.*

Results Alum turns the cabbage paper purple, cream of tartar turns the paper pink, and the Fruit Fresh produces a rose color.

Why? The amount of acid present determines the final color change. A strong acid will produce a red color. This test indicates that Fruit Fresh has the most concentration of acid, cream of tartar is next in concentration, and the alum has the least amount of acid. The purple color is produced by the combination of the blue in the test solution and the small amount of red caused by the acid properties in the alum.

93. Drinkable Acid

Purpose To identify a drinkable acid.

Materials *lemonade*
cabbage indicator (*prepared in experiment* CABBAGE INDICATOR)
glass
tablespoon

Procedure

■ *Place 1 tablespoon of cabbage indicator juice in the glass.*
■ *Add 1 tablespoon of water*
■ *Add 1 tablespoon of lemonade and stir.*

Results The blue indicator solution turns red.

Why? Lemonade, like the juice of all citrus fruits, contains citric acid. The blue cabbage solution turns red when mixed with an acid.

CABBAGE
JUICE

94. Baking with Acid?

Purpose To observe the effect that an acid has on baking.

Materials *vinegar*
6 cups
2 teaspoons
2 tablespoons
baking powder
baking soda
2 sheets of paper

Procedure

■ *Fill one cup one-half full with vinegar.*

■ *Fill another cup with water.*

■ *Separate the two sheets of paper and lay them on a table.*

■ *Place 2 cups on each sheet of paper.*

■ *Put 1 teaspoon of baking powder in two of the cups sitting on one of the sheets of paper. Write* BAKING POWDER *on the paper, and #1 in front of one cup and #2 in front of the other cup.*

■ *Use a clean teaspoon to place 1 teaspoon of baking soda in the remaining two cups. Write* BAKING SODA *on the paper. Number the cups #3 and #4.*

■ *Start with the baking power cups. Add 2 tablespoons of water to cup #1. Add 2 tablespoons of vinegar to cup #2.*

■ *Observe the results. It is always best to write down observations. Use the paper the cups are sitting on to record the results.*

■ *Add 2 tablespoons of water to cup #3 which contains baking soda.*

■ *Add 2 tablespoons of vinegar to cup #4 that also contains baking soda.*

■ *Observe and record the results.*

Results Foam is produced in cups #1, #2, and #4 when the liquid is added. Cup #3 only makes a thick milky-looking solution.

Why? Baking powder is a mixture of sodium bicarbonate, acid, and other materials. Water activates the powdered acid. The activated acid reacts with the sodium bicarbonate to produce carbon dioxide gas. Vinegar is an acid, and, like all acids, it reacts with sodium bicarbonate to produce carbon dioxide gas. It is the carbon dioxide gas that is needed to make a cake or bread rise during baking. The carbon dioxide pushes the batter up and heat bakes it in this elevated position.

Baking soda contains sodium bicarbonate and it will only produce carbon dioxide when combined with an acid. An acid would have to be added to a batter if baking soda was used as the source of carbon dioxide. Vinegar, cream of tartar, and buttermilk are all used as a source of acid. Anyone of these substances could be used with baking soda to produce carbon dioxide gas.

95. Turmeric Paper

Purpose To make a testing paper that will indicate the presence of a base.

Materials *zip-lock plastic bag*
teaspoon
⅓ cup alcohol
¼ teaspoon turmeric powder
coffee filters
cup
cookie sheet
quart bowl

Procedure

■ *Fill a cup one-third full with alcohol.*
■ *Stir ¼ teaspoon powdered turmeric into the alcohol.*
■ *Pour the solution into the quart bowl.*
■ *Dip one coffee filter at a time in the turmeric solution.*
■ *Place each wet filter on the cookie sheet and allow them to dry.*
■ *Cut the dry papers into strips about one-half inch by three inches.*
■ *Store the strips in a zip-lock plastic bag.*

Results The dry turmeric paper is a bright yellow.

Why? *Indicators* are materials that have a specific color change. Turmeric is an indicator for a base. The color change is from yellow to red.

ALCOHOL
COFFEE
FILTERS
TURMERIC

96. Now It's Red!

Purpose To produce a color change with an invisible gas.

Materials *turmeric paper* (*prepared in experiment* TURMERIC PAPER)
household ammonia

Procedure

■ *Moisten one end of a piece of turmeric paper with water.*

■ *Open the bottle of ammonia.*

Important: DO NOT *inhale the escaping fumes.*

■ *Hold the moistened paper about two inches above the open bottle. Do not touch the bottle with the papers.*

Results The wet end of the paper turns red.

Why? Household ammonia is a solution of ammonia gas dissolved in water. The smell observed when the bottle is opened is the escaping ammonia gas. This escaping gas mixes with the water on the paper to form the basic ammonia solution which turns the turmeric paper red.

Ginger
AMMONIA

97. Wet Only

Purpose To observe that dry solids must be wet to be tested with turmeric paper.

Materials *turmeric testing paper* (*prepared in experiment* TURMERIC PAPER)
baking soda
cup
teaspoon

Procedure

■ *Place ½ teaspoon of baking soda in the cup.*

■ *Touch the dry powder with a dry piece of turmeric paper.*

■ *Wet one end of the turmeric paper, touch the baking soda with the wet end.*

Results There is no change when the dry paper is used. The wet paper turns red.

Why? Baking soda is basic, but it must be dissolved in water before it can react with the colored chemicals on the turmeric paper. The water allows the chemicals to mix together.

Brenda
BAKING
SODA

98. Basic Cleaners

Purpose To test for the presence of a base in common cleaners.

Materials *12-inch sheet of aluminum foil*
teaspoon
5 turmeric testing strips
cup of water
lava soap
glass cleaner
oven cleaner
powdered abrasive cleaner

Procedure

■ *Lay the sheet of aluminum foil on a table.*

■ *Place ½ teaspoon of each of the four cleaners on the aluminum foil. Space them so that they do not touch.*

■ *Dip the end of one turmeric strip in the water. Lay the wet end on one of the testing materials.*

■ *Continue to wet the turmeric strips until one is placed on top of each of the four materials to be tested.*

Results All four of the strips turn red where they touch the materials.

Why? Many cleaners are basic. This is because bases combine with grease to form soap. The cleanser reacts with the unwanted grease and the soap that is formed is washed away.

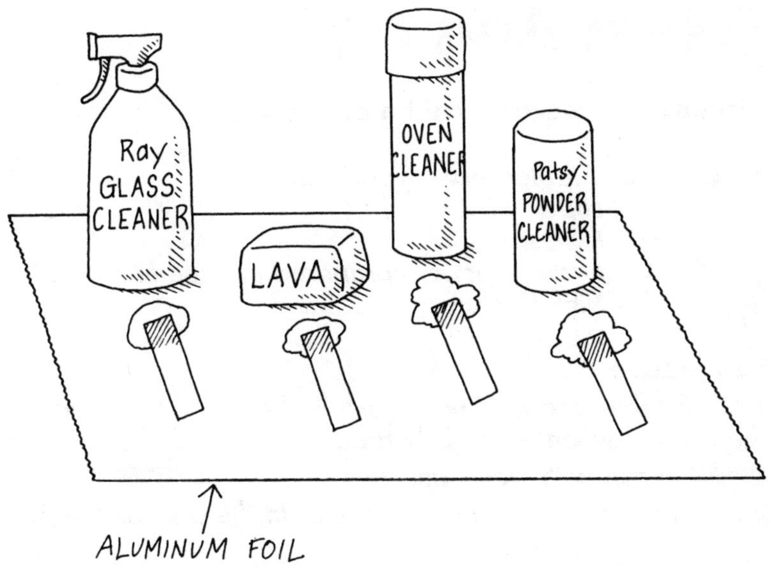
Ray
GLASS
CLEANER
LAVA
OVEN
CLEANER
Patsy
POWDER
CLEANER
ALUMINUM FOIL

99. Wood Ash

Purpose To make and test a basic solution.

Materials *2 tablespoons wood ash*
cup
tablespoon
turmeric paper (prepared in experiment TURMERIC PAPER)

Procedure

- *Put 2 tablespoons of wood ash in the cup. Wood ash is the ash left when wood is burned.*
- *Fill the cup with water and stir.*
- *Dip one end of the turmeric paper in the ash solution.*

Results The yellow paper turns red.

Why? Wood ash contains a chemical called potash. Potash is basic, and turmeric paper turns red when dipped into basic solutions.

100. **Neutral**

Purpose To neutralize a basic solution.

Materials *turmeric paper* (*prepared in experiment* TURMERIC PAPER)
household ammonia
vinegar
2 eyedroppers

Procedure

■ *Dip one end of the turmeric paper into the ammonia.*

■ *Fill the eyedropper with vinegar.*

■ *Drop vinegar on the end of the turmeric paper that is wet with the ammonia.*

Results Ammonia turns the turmeric paper red. The vinegar drops change the red color of the paper back to yellow.

Why? Ammonia is a base and vinegar is an acid. The combination of an acid and a base cancel each other. The products formed are not acidic or basic. The basic ammonia causes the turmeric paper to turn red. The drops of vinegar remove the basic ammonia by changing it to a non-basic chemical. Removing the ammonia changes the paper back to its original yellow color.

Ginger
AMMONIA
Davin
VINEGAR

101. Dissolving Fibers

Purpose To dissolve hair in bleach.

Materials *piece of hair (about the size of a walnut)*
bleach
small baby food jar
teaspoon

Warning: Do this experiment with adult supervision. If the bleach is spilled, immediately rinse the bleach off with water.

Procedure

■ *Fill the jar one-quarter full with bleach.*

■ *Collect a small sample of hair from a local beauty or barber salon. Place the hair in the jar with the bleach.*

■ *Use the spoon to push the hair down into the bleach so that the fibers become wet.*

■ *Allow the jar to set undisturbed for about 20 minutes.*

Results Foam forms on the surface of the bleach and small bubbles are seen on the hair. The hair is partially or completely dissolved.

Why? Bleach is a basic chemical and hair is an acid. The combination of an acid and a base is called a *neutralization reaction.* The materials produced by a neutralization reaction are totally different from the acid and base that were mixed. Bleach can dissolve any fiber that has acidic properties. Bleach is safe to use on cotton because cotton is basic, but it will dissolve acidic wool.

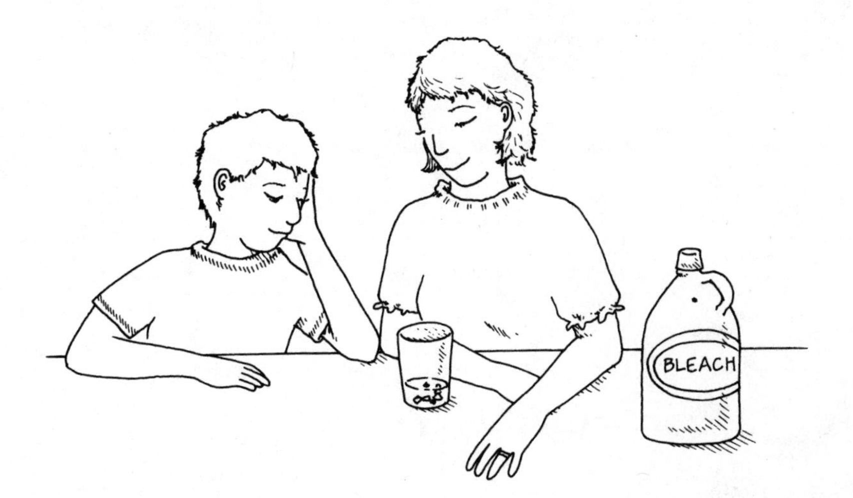
BLEACH

Glossary

Acid. A material that tastes sour, neutralizes bases, and turns purple cabbage juice red.

Atom. The smallest part of an element. It contains a positive center with negative charges spinning around the outside.

Base. A material that tastes bitter, neutralizes acids and turns purple cabbage juice green and turns turmeric paper red.

Capillary Action. The movement of a liquid in a thin tube due to the differences in pressure inside and outside the tube.

Catalase. An enzyme found in living cells.

Colloid. A solution containing tiny undissolved particles that permanently remain suspended in the liquid.

Concentrated. Pure, not weakened by adding other materials.

Contract. To become smaller by drawing closer together.

Deduction. A type of reasoning in which a conclusion is formed from the experimental results.

Density. The scientific way of comparing the "heaviness" of materials. It is a measurement of the mass of a specific volume.

Diffusion. The movement of molecules from one place to another, resulting in an even distribution of the molecule particles.

Dilute. To lessen the strength by mixing with something else, usually water.

Effervescence. Is created by dissolving a gas in a liquid and then adding a soluble solid.

Electron. Negative particle that spins around the nucleus of an atom.

Enzyme. A chemical found in living cells that changes the speed of the chemical reaction in the cell.

Evaporation. The changing of a liquid to a gas by increasing the heat content of the liquid.

Expand. To spread out; to get larger.

Freeze. To change a liquid to a solid by reducing the heat content of the liquid.

Gravity. The force that pulls objects on the earth toward the center of the earth.

Hydrogen Bond. A weak attraction between the hydrogen atom on one molecule with a hydrogen atom on another molecule. The attraction between hydrogen atoms on two water molecules is an example of hydrogen bonding.

Immiscible. The inability of two liquids to mix.

Inertia. The property of a material that resists any change in its state of rest or motion.

Matter. The substance things are made of. Matter takes up space and has inertia and mass.

Molecules. The linking of two or more atoms produces a molecule.

Neutralization. A process in which an acidic or basic solution is brought to a neutral state, one which is neither acidic nor basic.

Porous. Full of holes, hence, able to absorb liquids.

Reflective. To bounce back from a surface.

Saturated. When no additional solute can dissolve in a solvent.

Solute. The material that breaks into smaller parts and moves throughout a solvent.

Solvent. The material that a solute dissolves in.

Starch. A large molecule found in living cells. It combines with iodine to form a distinctive blue-black color.

Suspension. A mixture of two materials; one does not dissolve in the other, but temporarily stays suspended in the liquid until gravity pulls it down.

Tyndall Effect. Reflection of light by particles suspended in a solvent.

Vacuum. A space empty of matter.

Volume. Space occupied by matter.

Resource List

Amery, Heather, and Littler, Angela. *The FunCraft Book of Magnets and Batteries.* New York: Scholastic Book Service, 1976.

Armstrong, H. A., and Newbury, N. F. *The Young Experimenter.* New York: Sterling Publishing Co., 1960.

Cobb, Vicki. *Science Experiments You Can Eat.* New York: J. B. Lippincott, 1972.

Cobb, Vicki. *More Science Experiments You Can Eat.* New York: J. B. Lippincott, 1979.

Editors of the *Young People's Science Encyclopedia. Young People's Science Dictionary.* Chicago: Children's Press, Inc., 1964.

Herbert, Don. *Mr. Wizard's Supermarket Science.* New York: Random House, 1980.

Levenson, Elaine. *Teaching Children About Science: Ideas and Activities that Every Teacher and Parent Can Use.* Englewood Cliffs, N.J.: Prentice-Hall, Inc., 1985.

Lynde, Carleton John. *Science Experiments with Home Equipment.* Princeton, N.J.: C. Van Nostrand Co., Inc., 1949.

VanCleave, Janice Pratt. *Teaching the Fun of Physics: 101 Activities to Make Science Education Easy and Enjoyable.* Englewood Cliffs, N.J.: Prentice-Hall, Inc., 1985.

Index

acetic acid, 92. *See also* vinegar
acids, 189–217
  in Alka-Seltzer, 78
  in baking, 202–3
  bases combined with, 214
  definition of, 219
  drinkable, identification of, 200–201
  testing for, 190–91, 192–3, 194–95, 196–97
  testing for relative strength of, 198–99
  See also vinegar
air
  as example of matter, 14
  in rusting, 89
  space taken up by, 22
  See also gases
alchemists, xvii–xviii
alcohol
  combined with water, 26–27
  evaporation of, 130–31
  floating on water, 58–59
  perfume-making with, 172–73
  pulling power of, 40–41, 42
Alka-Seltzer, 78–79
alum, 98–99, 198–99
aluminum foil, absorption of light by, 186–87
aluminum hydroxide, 98
ammonium acetate, 102
ammonium hydroxide, 98, 100, 102
atoms
  constant motion of, 12
  definition of, 219
  charges of, 8, 10

baking powder, acid in, 202–3
baking soda
CO_2 produced by acid and, 76, 78, 202–3
turmeric paper testing of, 208
balloon experiments, 8–9, 10–11, 20–21, 28–29, 118–19
bases, 189–217
acids combined with, 214
cabbage indicator testing for, 190–91
cabbage paper testing for, 192–93, 194–95, 196–97
definition of, 219
neutralization of, 214–15
turmeric paper testing for, 204–5, 206–7, 208–9, 210–11, 212–13
black objects, absorption of light by, 186–87
bleach
decolorization by, 82–83, 84–85
dissolving hair in, 216
heat from water and, 184
boric acid crystals, miniature snow storm from, 156–57
breath
temperature of, 130
testing for CO_2 in, 72–73

cabbage indicator
in identifying drinkable acids, 200–201
making of, 190–91
in testing for relative strengths of acids, 198–99
cabbage paper
making of, 192–93
in testing for acids or bases, 194–95, 196–97
calcium carbonate, 94, 116
candy, dissolving of, 150–51
capillary action, 36
definition of, 219
carbon dioxide (CO_2)
from Alka-Seltzer, 78
cork popped by, 68–69
in effervescence, 66
in escaping soda bubbles, 64
in exhaled breath, 72
formed by vinegar and egg shell, 94
as necessary in baking, 202–3
produced by yeast, 68, 74–75
testing solution for, 70
catalase, 96
definition of, 219
celery, rising water in, 36–37
changes, 91–119
chemical reactions
gels from, 98–99, 102–3
heat from, 182, 184
milky, magnesia solution formed, 100–101
in tests for iron in fruit juices, 112–13
vinegar-baking soda, 118–19
in your mouth, 108
chemistry
history of, xvii–xviii
nature of, xvii
chewing, as part of chemical reaction, 108

cleaners, testing for bases in, 210–11
clicking coin experiment, 180–181
clothes, color of, 186
CO_2. *See* carbon dioxide
colloid
definition of, 219
milk as example of, 114
color
and absorption of light, 186–87
of diluted and concentrated solutions, 160–61
disappearing, 82–83
fading, 84–85
separation of, in ink, 154–55
streamers of, 148–49
concentrated, definition of, 160, 219
contract, definition of, 131, 219
control, definition of, 89
copper acetate, 92
cork, popping a, 68–69
cotton, bleach safe to use on, 216
cream of tartar, testing of, 198–99
crystals
boric acid, miniature snow storm from, 156–57
cubic salt, 142
dissolving of, 148–49
fluffy white, 134–35
lacy salt, 140–41
message written with, 132–33
needle-shaped, of epsom salt, 138
curd, 114
decant, definition of, 70
deduction, definition of, 219
deductive reasoning, 4
density
changing of, 28
definition of, 219
detergent, 38, 54–55
diffusion, 12
definition of, 220
dilute, definition of, 160, 220
dilution, color changes from, 160–61, 170–71
dissolving
of candy in mouth, 150–51
definition of, 148
of fiber by bleach, 216
heat and, 152
immiscible liquids and, 168
of ink into separate colors, 154–55
into oil and water, 158–59, 168–69
of powdered fruit drink, 148–49
saturated solution and, 156

effervescence
definition of, 220
in salt-soda experiment, 66–67
egg, vinegar to remove shell of, 94–95
electrons, 8, 10
definition of, 220
emulsions, definition of, 168
enzymes
definition of, 220
fruit discoloration caused by, 80

enzyme (*continued*)
hydrogen peroxide broken apart by, 96
epsom salt
crystals of, 138
milky solution from, 100–101
evaporation
of alcohol on thermometer bulb, 130–31
crystals formed by, 132–33, 134, 140–41, 142
definition of, 132, 220
expand, definition of, 130, 220

fibers, dissolving of, in bleach, 216
floating egg, 30–31
forces, 33–61
freeze, definition of, 220
freezing
of water molecules, 124–25
of orange juice, 126–27
salt and lower temperature for, 128, 136
frosty can, 136–37
fruit
cabbage paper testing of, 196–97
discoloration of, prevented by vitamin C, 80–81
Fruit Fresh, testing of, 198–99
fruit juices, testing for iron in, 112–13

gases, 63–89
from chemical reaction, 118
cooling and heating of, 180
space taken up by, 22
See also air
gels, produced by chemical reactions, 98–99, 102–3
gold, production of, xvii–xviii
gravity, 2
on bodies in liquid, 58
definition of, 220
overcoming force of, 8–9, 44–45
surface tension and, 42, 44
grease, combining of bases with, in cleaners, 210–11
green blob, 102–3
gypsum, 144

H_2O. *See* water
hair, dissolving of, in bleach, 216
heat, 175–87
from chemical reaction, 182, 184
dissolving speeded by, 152
evaporation speeded by, 132
from phase change of Plaster of Paris, 144
removed from water by salt, 122
rise of thermometer liquid through, 130–31
household ammonia
cabbage paper testing of, 194–95, 196–97
crystals from other chemicals and, 134–35
in gel experiment, 98
green blob from iron acetate and, 102–3
milky substance from epsom salt and, 100–101
neutralization of, 214

turmeric paper test of, 206–7
hydrogen bond, definition of, 220
hydrogen peroxide, breaking apart of, 96

ice, more space taken up by, than by water, 124–25
identifying unseen objects, 4–5
immiscible, definition of, 158, 168, 220
indicators, definition of, 204–5
inertia
in coin experiment, 2–3
definition of, 220
ink, separation of colors in, 154–55
iodine
in magic writing, 110
in testing for starch, 104–5, 106–7, 108
iron, testing for, 112–13
iron acetate, 102–3
iron hydroxide, 102
iron oxide, 89

layering
of oil and water, 158–59
of undissolved materials, 164–65
lemon juice, in magic writing, 110–11
light
color and absorption of, 186–87
newspaper aged by, 86–87
reflection of, in suspensions, 166–67
limestone
chemical removal of, 116–17
formation of, 72, 116
limewater
in limestone formation, 72, 116
making of, 70–71
in testing breath for CO_2, 72–73
in yeast experiment, 74–75

magnesium hydroxide, 100
magnesium sulfate, 100
matter, 1–31
air as example of, 14
contracted by cooling, 180
definition of, 2, 220
inertia as property of, 2–3
pieces of, cannot occupy same space at same time, 16, 18, 20, 22, 24, 66
producing different form of, 118–19
measuring jar, 24–25
milk, separation of, into component parts, 114–15
Milk of Magnesia, 100
molecules
attraction between, 56–57
definition of, 220
motion of, 12
moving together of, 20
pressure of, 14
of water. *See* Water
motion
molecular, 12
without touching, 10–11
unseen, 12–13

neutralization, 214
  definition of, 220
newspaper, rapid aging of, 86–87

oil
  spheres of, 58–59
  water and, as immiscible, 158, 168
opaque, definition of, 70
orange juice, frozen, 126–27
oxygen
  from bleach, 82, 84
  enzyme reaction with, 80
  in newspaper, 86
  in rusting, 89

paper hop, 8–9
pennies, green, 92–93
perfume, making, 172–73
phase changes, 121–45
photosynthesis, xix
physical properties, changes of, 6–7
plants
  photosynthesis by, xix
  rising water in, 36–37
  in yeast, 68, 74
Plaster of Paris, 144
porous, definition of, 220
potash, 212
potato, in breakdown of hydrogen peroxide, 96–97
precipitate, definition of, 70, 148

radiation, color and absorption of, 186–87
reasoning, deductive, 4
reflection of light in suspensions, 166–67
reflective, definition of, 220
rice, grains of, 16
rusting
  heat from, 182
  prevention of, 88–89

saliva
  chemical reaction from, 108
  in dissolving candy, 150
salt
  colder water from, 122–23
  cubic crystals from, 142
  freezing temperature of water lowered by, 128, 136
  lacy crystals of, 140–41
  shiny crystal message from, 132–33
  See also epsom salt
saturated, definition of, 156, 220
settling, 16, 162
shaking an emulsion, 168–69
snow storm, miniature, 156–57
soap, formed from a base and grease, 210
soap bubbles, 60–61
soda
  escaping bubbles from, 64–65, 66–67
  foamy, 66–67
soda bottle, shooting cork from, 68–69
sodium hypochlorite. *See* bleach
solute
  definition of, 221
  dissolving of, in solvent, 148–

49, 150–51, 152–53, 160–61
solutions, 147–73
saturated, 156, 220
solvent
definition of, 221
dissolving of solute in, 148–49, 150–51, 152–53, 160–61
space
between grains of matter, 16, 24
between water molecules, 26, 124
spheres, 58–59
spinning, parts of suspension cleared by, 162–63
starch
definition of, 221
testing for, 104–5, 106–7, 108–9
steel wool
iron acetate from, 102–3
in rusting experiment, 88–89
stirring, dissolving speeded up by, 150, 152
strengths of tea, 160–61
sugar
from chewing bread, 108
grains of, 24
sunlight, newspaper aged by, 86–87
super chain, 6–7
surface tension, gravity's effect on, 42, 44
suspensions
definition of, 164, 221
layering in, 164–65
parts of, separated by spinning, 162–63
reflection of light in, 166–67

talcum powder, 54–55
tea
comparing strengths of, 160–61
in tests of fruit juices for iron, 112–13
temperature. *See* heat
thermometer, rise and fall in liquid of, 130–31
turmeric paper
making of, 204–5
in neutralizing a base, 214–15
testing ammonia with, 206–7
in testing for bases in cleaners, 210–11
testing wood ash with, 212–13
wetness needed for tests with, 208
Tyndall Effect, definition of, 167, 221

unseen object, identifying, 4–5

vacuum, definition of, 221
vinegar
cabbage paper testing of, 194–95, 196–97
chemical name of, 92
CO_2 produced by baking soda and, 76
gas from baking powder and, 202–3
gas from baking soda and,

vinegar (*continued*)
118, 202–3
iron acetate from steel wool and, 102–3
limestone dissolved by, 116–17
in milk separation, 114
neutralization of, 214
pennies colored green by, 92–93
in removing shell of egg, 94–95
in rusting experiment, 88–89
vitamin C
fruit discoloration prevented by, 80–81
in magic writing, 110
volcano, simulation of eruption of, 76–77
volume, definition of, 221

water
attraction of molecules of, 34–35, 44, 46, 48–49, 50, 52
boiling appearance of, 34–35
cannot occupy space with other matter, 18–19
clearing mixture of flour and, 162–63
cold and warm, difference in weight of, 176–77, 178–79
dilution within, 170–71
disappearing color in, 82–83, 84–85
above edge of cup, 46–47, 48–49
expansion of, when frozen, 124–25
freezing temperature of, 128
heat speeds up molecules of, 152
lowering temperature of, 122–23
oil and, as immiscible, 158, 168
phase change from addition of, 144
pulling power of molecules of, 38, 40–41
rising, into plants, 36–37
salted, floating egg in, 30–31
salted, lower freezing temperature of, 128, 136–37
smoke-like rings of color in, 176–77, 178–79
soap bubbles and, 60–61
spaces between molecules of, 26, 124
suspensions in, 162–63, 164–65
wetting agents, 54–55
whey, 114
wood ash, 212
wool, bleach not safe to use on, 216
writing
with lemon and iodine, 110–11
with shiny crystals, 132–33

yeast, carbon dioxide produced by, 68, 74–75